SpringerBriefs in Applied Sciences and Technology

Thermal Engineering and Applied Science

Series Editor

Francis A. Kulacki, Department of Mechanical Engineering, University of Minnesota, Minneapolis, MN, USA

More information about this series at http://www.springer.com/series/8884

Sujoy Kumar Saha • Hrishiraj Ranjan
Madhu Sruthi Emani • Anand Kumar Bharti

Heat Transfer Enhancement in Externally Finned Tubes and Internally Finned Tubes and Annuli

Springer

Sujoy Kumar Saha
Mechanical Engineering Department
Indian Institute of Engineering
Science and Technology, Shibpur
Howrah, West Bengal, India

Hrishiraj Ranjan
Mechanical Engineering Department
Indian Institute of Engineering
Science and Technology, Shibpur
Howrah, West Bengal, India

Madhu Sruthi Emani
Mechanical Engineering Department
Indian Institute of Engineering
Science and Technology, Shibpur
Howrah, West Bengal, India

Anand Kumar Bharti
Mechanical Engineering Department
Indian Institute of Engineering
Science and Technology, Shibpur
Howrah, West Bengal, India

ISSN 2191-530X   ISSN 2191-5318   (electronic)
SpringerBriefs in Applied Sciences and Technology
ISSN 2193-2530   ISSN 2193-2549   (electronic)
SpringerBriefs in Thermal Engineering and Applied Science
ISBN 978-3-030-20747-2   ISBN 978-3-030-20748-9   (eBook)
https://doi.org/10.1007/978-3-030-20748-9

This Springer imprint is published by the registered company Springer Nature Switzerland AG
The registered company address is: Gewerbestrasse 11, 6330 Cham, Switzerland

Contents

1 Introduction 1
References 6

2 Round Tubes Having Plain-Plate Fins 7
References 25

3 Circular Fins with Staggered Tubes, Low Integral Fin Tubes 31
References 37

**4 Enhanced Plate Fin Geometries with Round Tubes and Enhanced
Circular Fin Geometries** 39
References 61

**5 Oval and Flat Tube Geometries, Row Effects in Tube Banks,
Local Heat Transfer Coefficient on Plain Fins, Performance
Comparison, Numerical Simulation and Patents, Coatings** 69
5.1 Oval and Flat Tube Geometries 69
5.2 Row Effects in Tube Banks 74
5.3 Local Heat Transfer Coefficient on Plain Fins 74
5.4 Performance Comparison, Numerical Simulation
and Patents, Coatings 77
References 81

6 Internally Finned Tubes and Spirally Fluted Tubes 85
References 118

7 Advanced Internal Fin Geometries and Finned Annuli 127
References 157

8 Conclusions 163

Additional References 165

Index 171

Nomenclature

Δp	Air-side pressure drop, Pa or lbf/ft^2
Δp_f	Pressure drop assignable to fin area in finned tube exchanger, Pa or lbf/ft^2
a	Major diameter for rectangular tube cross section, m or ft
A	Total heat transfer surface area (both primary and secondary, if any) on one side of a direct transfer type exchanger; total heat transfer surface area of a regenerator, m^2 or ft^2
A_c	Flow cross-sectional area in minimum flow area, m^2 or ft^2
A_f	Fin or extended surface area on one side of the exchanger, m^2 or ft^2
A_{fa}	Actual flow area of an internally finned tube, $A_n(1-2e/d_i)^2$, m^2 or ft^2
A_{fin}	Inter-fin flow area of an internally finned tube, $A_{fa}-A_{core}$, m^2 or ft^2
A_{fr}	Heat exchanger frontal area, m^2 or ft^2
A_n	Nominal flow area of an internally finned tube $\Pi d_i^2/4$, m^2 or ft^2
AMTD	Arithmetic mean temperature difference K
B	Minor diameter for rectangular tube cross section, m or ft
D_{ab}	Diffusion coefficient for component a through component b, m^2/s
D_i	Internal diameter, m or ft
d_o	Outside diameter, m or ft
D_{vi}	Volume-based tube inner diameter, m or ft
e/d	Rib height, dimensionless
f	Fanning friction factor $\Delta P\rho D_h/2LG^2$, dimensionless
f_f	Friction factor of fins in Eq. 6.5 ($=2\,\Delta p \pm \rho A_c/A_j G^2$), dimensionless
f_t	Friction factor of tubes in Eq. 6.5 ($=2\,\Delta p \pm \rho A_c/A_t G^2$), where $A=A-A_f$, dimensionless
f_{tb}	Tube bank friction factor $=\Delta P\rho D_h/2NG^2$, dimensionless
Gr	Grashof number $=g\beta\Delta TD_h^3/v^2$, dimensionless
Gz	Graetz number $=\Pi d_i RePr/4L$, dimensionless
j	$StPr^{2/3}$ dimensionless
n_L	Number of louvers in airflow depth, dimensionless
Nu_{Dh}	Nusselt number $=hD_h/k$, dimensionless
p/d	Rib pitch, dimensionless

p_f	Fin pitch, center to center spacing, m or ft
p_w	Wave pitch of wavy fin, m or ft
Re_{Dvi}	Reynolds number based on D_{vi}, GD_{vi}/μ dimensionless
Re_{Dvo}	Reynolds number based on D_{vo}, GD_{vc}/μ dimensionless
s	Spacing between two fins $= p_f - t$, m or ft
S_f	Flow frontal area of heat exchanger, m^2 or ft^2
Sh	Sherwood number for mass transfer $(=K_m D_h/D_{ab})$, dimensionless
St	Stanton number $= h/Gc_p$ dimensionless
$u*$	Friction velocity $= (\tau_u/\rho)^{1/2}$, m/s or ft/s
u_m	Fluid mean axial velocity at the minimum free flow area, m/s or ft/s
V_m	Heat exchanger tube material volume, m^3 or ft^3
w_{fin}	Weight of fins in heat exchanger, kg or lbm
w_s	Width of segmented fin (Fig. 6.20), m or ft
W_{tot}	Weight of tubes and fins in heat exchanger, kg or lbm
w_{tub}	Weight of tubes in heat exchanger, kg or lbm
X^+	$x/D_h RePr$, dimensionless
X_{DV}^+	$RePrD_v/L$, dimensionless

Greek Symbols

2θ	Included angle of fin cross section normal to flow, radians, or degrees
α	Helix angle relative to tube axis radians or degrees
β	Helix angle
γ	Reciprocal of fin pitch, m or ft
δ	Liquid film thickness
Δp	Pressure drop
ΔT	Temperature difference
ε	Permittivity
η_f	Fin efficiency or temperature effectiveness of the fin, dimensionless
η_o	Surface efficiency of finned surface $= 1 - (1 - \eta_f)A_f/A$, dimensionless
μ	Dynamic viscosity
θ	Louver angle for louver fin, radians
ρ	Density
σ	Surface tension
τ_w	Wall shear stress, Pa or lbf/ft^2

Subscripts

ave	Average
deff	Effective diameter
ev	Evaporation
fd	Fully developed flow

H1	Heat flux boundary condition
i	Inner
il	Inline tube arrangement
in	Inlet
L	Liquid
m	Average value over flow length
o	Outer
o	Outside (air-side) surface
p	Plain tube or surface
s	Saturated
st	Staggered tube arrangement
sub	Subcooled
v	Vapor
w	Evaluated at wall temperature
x	Local value

Chapter 1
Introduction

Finned tube heat exchangers are used for single-phase (for gases) or two-phase heat transport for liquids (both single and two phases) (Fig. 1.1). The fin surface may be plate fin-and-tube geometry or individually finned tubes. The tubes may be round type or oval or flat tubes. The arrangements of the tubes may be staggered tube arrangement or inline tube arrangement types. The fin may also be an integral fin tube used for liquids. Typically, lower fin heights are required for liquids than that for gases as liquids have higher heat transfer coefficients than gases.

It is necessary to increase air-side hA value, since the gas-side heat transfer coefficient is much smaller than the tube-side value. A plain surface geometry increases the air-side hA value by increasing the area A. Enhanced fin surface geometries provide higher heat transfer coefficients than a plain surface. Basic enhancement geometries are wavy and interrupted fins (Fig. 1.2). These enhancement techniques of Fig. 1.2 may be applied to circular tubes and flat extruded aluminium tubes. In these devices, pressure can be contained by the internal membranes existing in the devices.

Abbott et al. (1980) and Webb (1983, 1987) have studied this type of enhancement devices. O'Connor and Pasternak (1976) have used the fins slit from the thick wall of an aluminium extrusion and bent upwards (Shah and Webb 1982). These fins may be used mostly in domestic air-conditioning purpose, but these cannot be widely used in industry from the cost point of view. Nevertheless, some of these designs have now been introduced for use in automotive air-conditioning evaporators and condensers. Figure 1.3 shows a recent fin design where winglet-type vortex generators are formed radially.

High fin efficiency can be obtained if the fin material has high thermal conductivity. Operational constraints sometimes limit the fin density due to gas-side fouling. The fin material may be based on the operating temperature and corrosion potential. Aluminium is the most common material of the fins for residential and automotive air-conditioning, automotive radiators and process industry heat exchangers. Steel fins are used for boiler economizers and heat recovery exchangers.

© The Author(s), under exclusive license to Springer Nature Switzerland AG 2020
S. K. Saha et al., *Heat Transfer Enhancement in Externally Finned Tubes and Internally Finned Tubes and Annuli*, SpringerBriefs in Applied Sciences and Technology, https://doi.org/10.1007/978-3-030-20748-9_1

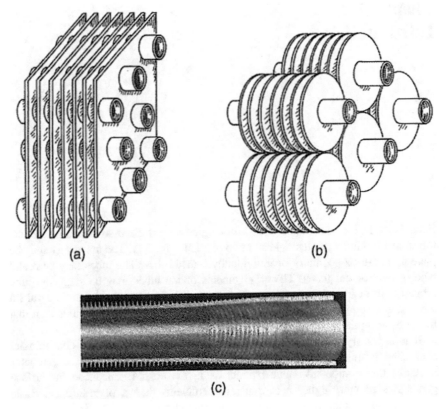

Fig. 1.1 Finned tube geometries used with circular tubes: (**a**) plate fin-and-tube used for gases, (**b**) individually finned tube having high fins, used for gases (from Webb 1987), (**c**) low, integral fin tube (Webb and Kim 2005)

In this research monograph, various fin geometries and their performance charac-teristics along with their alternatives are discussed, and these types of fin geometries are likely to yield the highest performance per unit heat exchanger core weight. Possible improvements in the air-side surface geometry are considered.

Internally finned tubes are mostly used for liquids and in some cases for pressur-ized gases, namely for in-tube gas flow in an air-compressor intercooler. Internally finned tubes for condensation and vaporization inside tubes have been discussed in other research monographs in this series (Two-Phase Heat Transfer Enhancement). Figure 1.4 shows internally finned tubes having integral internal fins made with axial or helical fins. The fin height for use with liquids will be limited due to efficiency concerns. Equations (1.1) and (1.2) express the fin efficiency.

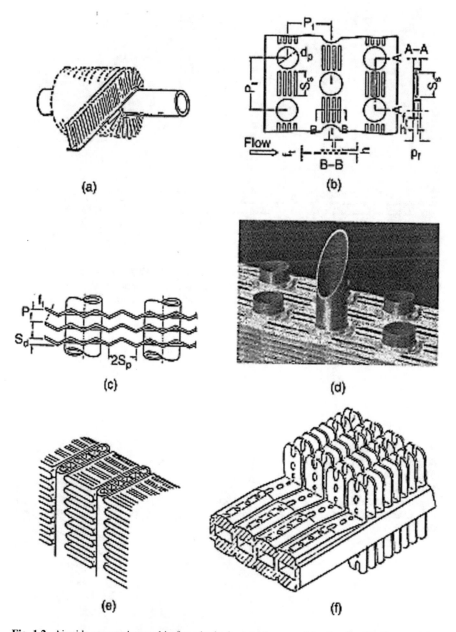

Fig. 1.2 Air-side geometries used in finned tube heat exchangers: (**a**) spine-fin, (**b**) slit-type OSF, (**c**) wavy fins, (**d**) convex louvre fin, (**e**) louvre fins brazed to extruded aluminium tube, (**f**) interrupted skive fin integral to extruded aluminium tube (Webb and Kim 2005)

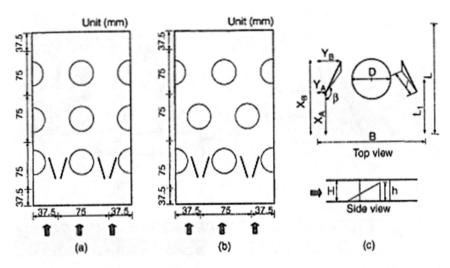

Fig. 1.3 Three-row fin-and-tube geometry tested by Torii et al. (2002): (**a**) in-line, (**b**) staggered arrangement, (**c**) shape of the vortex generator. Vortex generators are arranged in common flow-up configuration, $d = 30$ mm, $H = 5.6$ mm, $h = 5.0$ mm, 165° angle of attack (Torii et al. 2002)

Fig. 1.4 Illustration of integral fin tubes for liquids: (**a**) axial internal fins, (**b**) helical internal fins, (**c**) extruded aluminium insert device (Webb and Kim 2005)

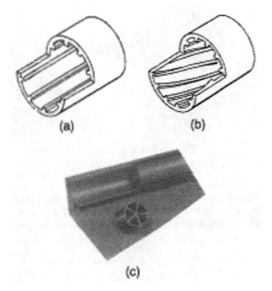

$$\eta_f = \frac{\tanh(me)}{me} \tag{1.1}$$

$$m = \left(\frac{2h}{k_f t}\right)^{1/2} \tag{1.2}$$

The insert device could be used for gases or liquids having a low heat transfer coefficient. The insert device must have good thermal contact with the tube wall.

The dynamics of flow in finned tube heat exchanger is very complex as it is. This is so because of the three-dimensional nature of the flow and flow separations. The fin geometries make the flow further complex. The numerical and analytical studies have made a dent in the fluid dynamics in finned tubes flow. However, much remains to be done. The predictive correlations, empirical in nature, have been developed by the experimental investigations, and these are based on power law correlations using multiple regression techniques. However, this requires a prior detailed knowledge of the geometric parameters of the fin and the flow variables involved. The variables are of several kinds:

(a) Flow variables are air velocity, viscosity, density, thermal conductivity and specific heat.
(b) Tube bank variables like tube root diameter, transverse tube pitch, row pitch, tube layout (if it is staggered or inline) and the number of rows.
(c) Fin geometry variables: fin pitch, fin height, fin thickness, in finned geometries additionally, wave height, wave pitch and wave shape.

It has to be noted here that no rules prevail for selecting the appropriate dimensionless geometric variables. Only trial basis may be adopted. Power law correlations are not any rationality based. They provide only empirical correlations of the data set. There is no point in extrapolating such correlations beyond the range of the variables used to develop the correlation.

One needs to identify a characteristic dimension that appears to dominate over the possible choices, and this dimension should be used to define Reynolds number. However, it must be noted here that the choice of characteristic dimension is arbitrary. The regime, laminar or turbulent, must be decided carefully. Eddies are shed in the tubes, and these wash over the fin surface and provide mixing of the flow.

Equation (1.3) defines the friction factor frequently used for tube banks (bare or finned).

$$f_{tb} N = \frac{fL}{D_h} \tag{1.3}$$

Kays and London (1984) and Rozenman (1976a, b) may be used for finding some basic data on enhanced surfaces.

References

Abbott RW, Norris RH, Spofford WA (1980) Compact heat exchangers in general electricproducts—sixty years of advances in design and in manufacturing technologies. In: Shah RK, McDonald CF, Howard CP (eds) Compact heat exchangers—history, technology, manufacturing technologies, ASME Symp, vol 10, pp 37–56

Kays WM, London AL (1984) Compact heat exchangers. McGraw-Hill, New York

O'Connor JM, Pasternak SF (1976) Method of making a heat exchanger, U.S. patent 3,947,941

Rozenman, T (1976a) Heat transfer and pressure drop characteristics of dry cooling tower extended surfaces, part I: heat transfer and pressure drop data, Report BNWL-PFR 7-100. Battelle Pacific Northwest Laboratories, Richland, WA, March 1

Rozenman, T (1976b) Heat transfer and pressure drop characteristics of dry cooling tower extended surfaces, part II: data analysis and correlation, Report BNWL-PFR 7-102. Battelle Pacific Northwest Laboratories, Richland, WA

Shah RK, Webb RL (1982) Compact and enhanced heat exchangers. In: Taborek J, Hewitt GF, Afgan N (eds) Heat exchangers: theory and practice. Hemisphere, Washington, DC, pp 425–468

Torii K, Kwak K, Nishino K (2002) Heat transfer enhancement and pressure drop for fin-tube bundles with winglet vortex generators. In: Heat transfer 2002. Proceedings of the 12th international heat transfer conference, vol 4. pp 165–170

Webb RL (1983) Heat transfer and friction characteristics for finned tubes having plain fins. In: Low Reynolds number flow heat exchangers. Hemisphere, Washington, DC, pp 431–450

Webb RL (1987) Enhancement of single-phase heat transfer. In: Kakac S, Shah RK, Aung W (eds) Hand book of single-phase heat transfer. Wiley, New York, pp 17.1–17.62

Webb RL, Kim NY (2005) Principles of enhanced heat transfer. Taylor & Francis, New York

Chapter 2
Round Tubes Having Plain-Plate Fins

Ohara and Koyama (2012) investigated the heat transfer and flow pattern experimentally in a plate-fin heat exchanger. They studied the thermo-hydraulic characteristics of falling film evaporation of pure refrigerant HCFC123 in a vertical rectangular channel with a serrated fin surface. The rear wall of the channel and evaporator was heated by electricity and liquid refrigerant flowing down vertically on it. The flow pattern was observed directly through transparent vinyl chloride resin plate during evaporation process. Experimental setup was supplied with constant mass velocity ($G = 28$–70 kg/m^{2}s), heat flux ($q = 20$–50 kWw/m^2) and pressure ($P = 100$ kPa). It was observed that when the vapour quality was more than equal to 0.3, heat transfer coefficient depended on the values of both mass velocity and heat flux. Figure 2.1 shows the variation of heat transfer coefficient with respect to vapour quality for heat fluxes of 20, 30, 40 and 50 kW/m^2, respectively. Figure 2.2 shows the variation of Nusselt number with respect to Reynolds number for mass velocity of 28, 40, 55 and 70 kg/m^{2}s, respectively.

Thermo-hydraulic characteristics in air-cooled compact wavy fin heat exchanger was reviewed and analysed by Awad and Muzychka (2011). They developed the new model to simplify the existing studies of Fanning friction factor f and the Colburn j factor. These models were developed by establishing the correlation between low Reynolds number and laminar boundary layer regions. The prepared model was based upon the geometrical and thermophysical parameters such as fin height (H), fin spacing (S), wave amplitude (A), fin wavelength (λ), Reynolds number (Re) and Prandtl number (Pr). They compared the proposed model with numerical and experimental data of published literature for air. Figure 2.3 shows characteristic dimensions and top view of a basic cell of wavy fin geometry. Muley et al. (2002, 2006), Muzychka (1999), Muzychka and Kenway (2009), Sheik Ismail et al. (2009, 2010), Zhang (2005), Zhang et al. (2003, 2004), Junqi et al. (2007), Rush et al. (1999) and Lin et al. (2002) also studied the effect of fins in heat exchanger.

© The Author(s), under exclusive license to Springer Nature Switzerland AG 2020
S. K. Saha et al., *Heat Transfer Enhancement in Externally Finned Tubes and
Internally Finned Tubes and Annuli*, SpringerBriefs in Applied Sciences and
Technology, https://doi.org/10.1007/978-3-030-20748-9_2

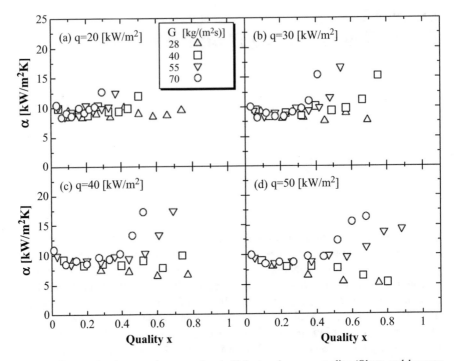

Fig. 2.1 The relation between heat transfer coefficient and vapour quality (Ohara and koyama 2012)

Bahrami et al. (2012) studied numerically and experimentally about the effect of geometric parameters of multi-louvred fins of compact heat exchangers on the heat transfer enhancement. They solved the three important thermophysical conservation equations of mass, momentum and energy using the finite volume method in various heat and flow conditions. Figure 2.4 shows the fin geometric parameters such as fin width (F_d) and fin pitch (F_p). Figure 2.5 shows the variation of pressure drop with inlet flow velocity or Reynolds number for the louvred angles of 18°, 24°, 26°, 28° 30° and 38° at the fin pitch of 1.3 mm. They observed that increasing the louvre angles resulted in more deviation of the fluid and that causes more pressure losses in the fins. They found from results that thermal capacity and pressure drop decreased with increase in fin pitch. Pressure drop and heat capacity were lesser affected by louvre angle than fin pitch.

They found that in the second row of louvres at low Reynolds number, more pressure loss and less heat transfer took place. Therefore, they recommended removing the second-half of louvres and studied the effect of this type of fin having 26° louvre angle. Figure 2.6 shows the semi-louvred fin. The results indicated that thermal heat capacity increased up to 22% in the semi-louvred fin at 2 m/s inlet velocity. Therefore, they strongly recommended this type of fin heat exchanger in stationary refrigeration and air-conditioning system. Atkinson et al. (1998), Chang and Hsu (2000), Chang and Wang (1996, 1997), Chang et al. (1994), Dillen and

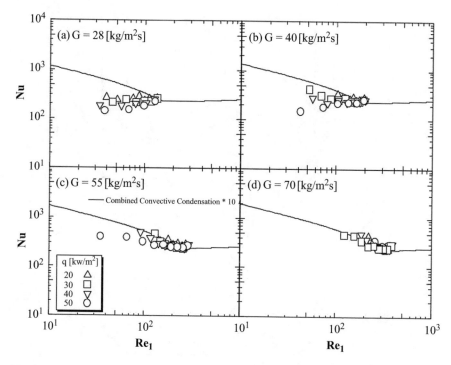

Fig. 2.2 The relation between Nusselt number and Reynolds number (Ohara and Koyama 2012)

Webb (1994), Dong et al. (2007), Lawson and Thole (2008), Lyman et al. (2002), Perrotin and Clodie (2004) and Sunden and Svantesson (1992) also studied the effect of louvred fin and tube heat exchanger on thermo-hydraulic characteristics.

Haghighi et al. (2018) conducted an experimental investigation in natural heat convection on thermal performance and convective heat transfer coefficient of plate fins and plate cubic pin-fin heat sink. He conducted the experimental investigation for Rayleigh number range of 8×10^6 to 9.5×10^6, heat input range of 10–120 W. Fin spacing and fin numbers were varied between 5 and12 mm and 5 and 9, respectively. They investigated the effect of fin spacing and number of fins of plate fins and plate cubic pin fin on thermal resistance and heat transfer. They found that plate cubic pin fin having 8.5 mm fin spacing and seven fins was better than plate-fin heat sink. They developed empirical correlations for average Nusselt number as a function of number of fin plates, fin spacing to height ratio as well as Rayleigh number.

Figure 2.7 shows the geometry of plate fin and plate cubic pin fin. Table 2.1 shows the dimensions of test fins. Figure 2.8 shows the variation of thermal resistance with fin spacing for plate pin fin and plate cubic pin fin. Figure 2.9 shows that Nusselt numbers were increasing with increase in Rayleigh numbers. The results of experimental investigation revealed that increasing the fin space caused lower thermal resistance but increase in fin number did not cause better heat transfer. They observed that thermal resistance of plate cubic pin fins were

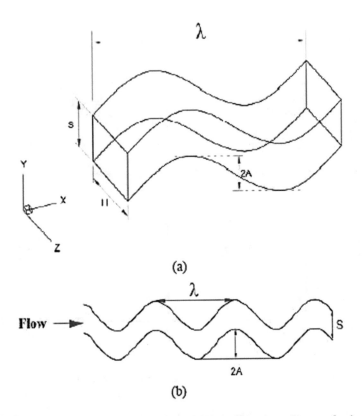

Fig. 2.3 Basic cell of wavy fin geometry: (a) characteristic dimensions of a wavy fin channel and (b) top view of a wavy channel (Awad and Muzychka 2011)

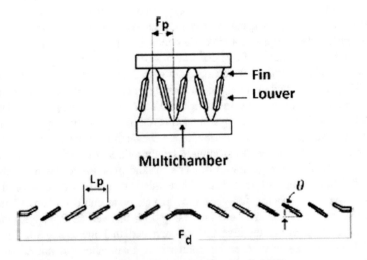

Fig. 2.4 Geometric parameters of fin and louvres (Bahrami et al. 2012)

Fig. 2.5 The variations of pressure drop vs. flow velocities for different louvre angles (Bahrami et al. 2012)

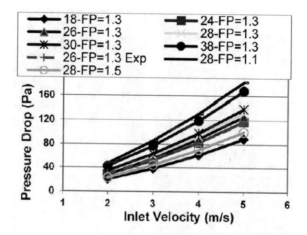

Fig. 2.6 Semi-louvred fin model (Bahrami et al. 2012)

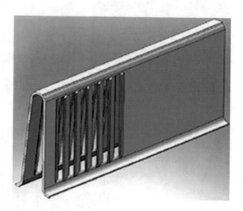

decreased by 15% compared to that of plate fin. Zaretabar et al. (2018), Mohammadian and Zhang (2017), Ji et al. (2018), Yang et al. (2017), Joo and Kim (2015), Yu et al. (2005), Yazicioğlu and Yüncü (2007), Yang and Peng (2009), Jeon and Byon (2017), Lee et al. (2016) and Micheli et al. (2016) investigated the effect of plate fins and pin fins on the hydrothermal characteristics.

Didarul Islam et al. (2008) reported the performance of rectangular fins having different patterns and placed in duct flow in different arrangements. Co-angular, zigzag, co-rotating and co-counter rotating configurations as shown in Fig. 2.10 have been used for the analysis. The friction factor variation with Reynolds number for the four configurations of fins, for different Prandtl numbers, has been shown in Fig. 2.11. The heat transfer coefficient correlations for different fin patterns and pitch ratios for a given fin height of 10 mm have been tabulated in Table 2.2. Also, the variation of η (ratio of Nusselt number of enhanced surface to the Nusselt number of plain surface) with $f^{1/3}$ Re has been plotted and presented in Fig. 2.12. They have observed that the friction factor for all four fin patterns were greater than that for the

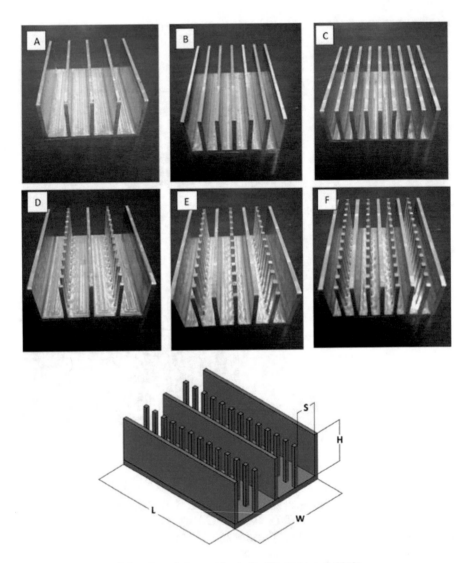

Fig. 2.7 Geometry of plate fin and plate cubic pin fin (Haghighi et al. 2018)

Table 2.1 Dimensions of test fins (Haghighi et al. 2018)

Fin type	Fin shape	Fin number	Fin spacing (mm)	S/H
Type A	Plate pin fin	5	12	24/90
Type B	Plate pin fin	7	8.5	17/90
Type C	Plate pin fin	9	5	10/90
Type D	Plate cubic pin fin	5	12	24/90
Type E	Plate cubic pin fin	7	8.5	17/90
Type F	Plate cubic pin fin	9	5	10/90

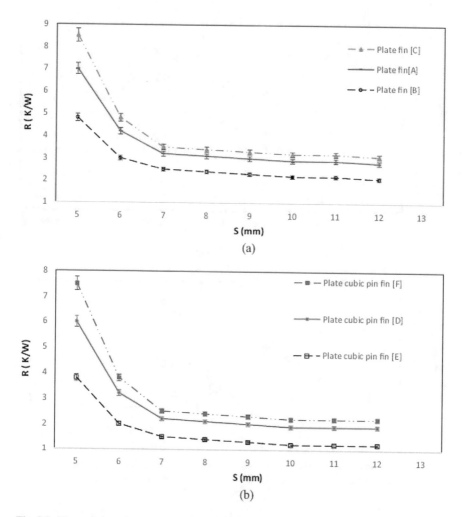

Fig. 2.8 The variation of thermal resistance with fin spacing. (**a**) Plate pin fin heat sinks. (**b**) Plate cubic pin fin heat sinks (Haghighi et al. 2018)

smooth rectangular duct for fully developed turbulent flow. The maximum pressure drop was seen in the duct with co-rotating fins.

This is because of strong flow interactions accompanied with vortex attack on the end wall and fin surface. The fins with co-angular pattern have showed minimum pressure drop. They have observed the smoke flow pattern around the fins and oil titanium oxide flow pattern on the end wall. They reported that they were both in good agreement. They have also observed that the flow over co-angular fin plate was governed by horseshoe vortices while wavy flow behaviour was dominant in the case of zigzag fin pattern. This is because the diverging fin pairs generate longitudinal vortices which attack the end wall and fin surfaces together. In case of

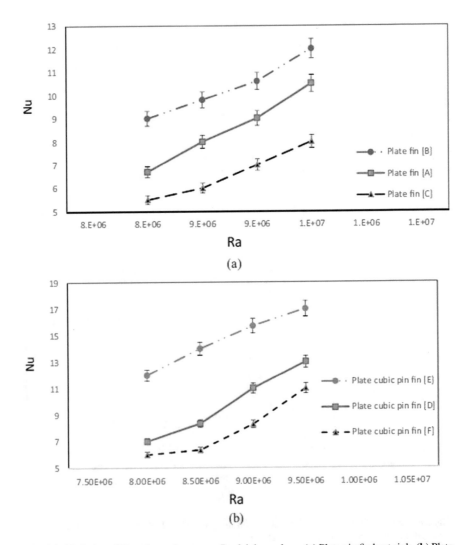

Fig. 2.9 Variation of Nusselt number versus Rayleigh numbers. (**a**) Plate pin fin heat sink. (**b**) Plate cubic pin fin heat sink (Haghighi et al. 2018)

co-counter rotating fin pattern, the flow was only slightly disturbed due to converging fin pairs. They concluded that the fin with co-rotating pattern with pitch ratio 2 and fin height 10 mm has shown the best thermal performance among all the fins considered. Also, the heat transfer using co-rotating fin pattern was noted to be threefold that of the duct without fins.

Torii and Yanagihara (1997) worked with vortex generators; Sparrow et al. (1982, 1983) studied rectangular fin arrays; and Kadle and Sparrow (1986), Turk and Junkhan (1986), Oyakawa et al. (1993), Molki et al. (1995), El-Saed et al. (2002) and Bilen and Yapici (2002) have carried out similar investigations for heat transfer

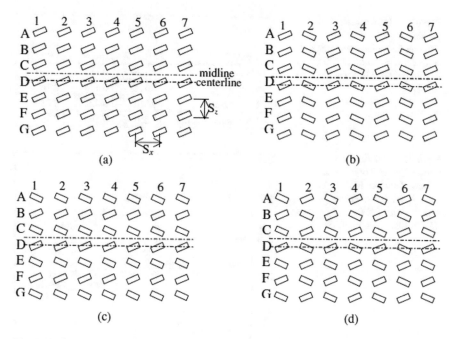

Fig. 2.10 Co-angular, zigzag, co-rotating and co-counter rotating configurations. (**a**) Co-angular pattern. (**b**) Zigzag pattern. (**c**) Co-rotating pattern. (**d**) Co-counter rotating pattern (Didarul Islam et al. 2008)

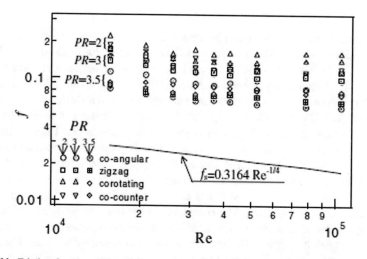

Fig. 2.11 Friction factor variation with Reynolds number for the four configurations of fins, for different Prandtl numbers (Didarul Islam et al. 2008)

Table 2.2 Heat transfer coefficient correlations for different fin patterns and pitch ratios for a given fin height of 10 mm (Didarul Islam et al. 2008)

	$\overline{Nu}_{overall} = cRe^{0.7}$		
	c		
Pattern	PR = 2	PR = 3	PR = 3.5
Co-angular	0.163	0.153	0.149
Zigzag	0.175	0.176	0.168
Co-rotating	0.261	0.223	0.212
Co-counter rotating	0.191	0.169	0.191

Fig. 2.12 Variation of η (ratio of Nusselt number of enhanced surface to the Nusselt number of plain surface) with $f^{1/3} Re$ (Didarul Islam et al. 2008)

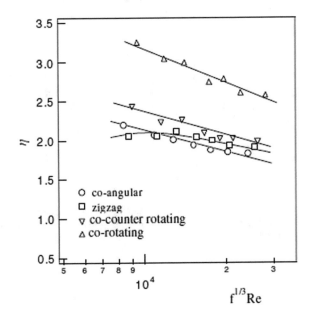

enhancement and pressure drop characteristics. Fabbri (1998, 1999), Zeitoun and Hegazy (2004), Olson (1992), Alam and Ghoshdastidar (2002), Saad et al. (1997), Kumar (1997), Yu et al. (1999), Liu and Jensen (1999), Sarkhi and Nada (2005), Wang et al. (2008a, b, c), Eckert and Irvine (1960), Yu and Tao (2004), Shih et al. (1995) and Park and Ligrani (2005) have carried out similar works.

Sajedi et al. (2015) worked on optimization of fin numbering in a heat exchanger having external extended finned tube for natural convection. They carried out numerical investigation and presented the results for heat transfer rate and average Nusselt number. The experiment was carried out for fixed Reynolds number and varying Rayleigh number. They have compared the experimental results with the numerical results and developed correlation for Nusselt number. The fin geometry has been shown in Fig. 2.13. The comparison of surface temperatures of the heat exchanger for numerical and experimental results has been presented in Fig. 2.14 for different Rayleigh numbers. The number of fins was considered to be 20. The rate of entropy generation $\left(\dot{S}_{gen} \right)$ and total heat loss (q) have been shown in Figs. 2.15 and 2.16, respectively.

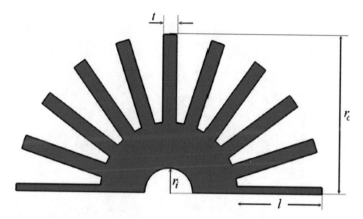

Fig. 2.13 Fin geometry (Sajedi et al. 2015)

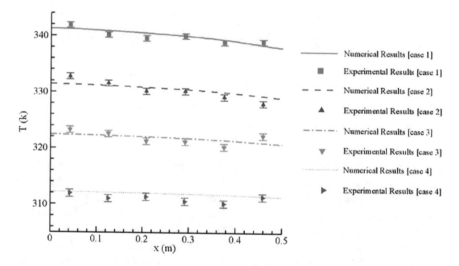

Fig. 2.14 Comparison of surface temperatures of the heat exchanger for numerical and experimental results (Sajedi et al. 2015)

The variation of average Nusselt number with the number of fins has been shown in Fig. 2.17. Three graphs have been shown to clearly present the average Nusselt number variation for different ranges of number of fins. They explained that as heat transfer surface increases and heat transfer coefficient decreases with the increase in number of fins, there is a definite need to obtain the optimum number of fins. For different cases considered for the analysis, the optimum number of fins ranged from 10 to 12.

Atayılmaz and Teke (2009, 2010), Ahmadi et al. (2014), Taghilou et al. (2014), Park et al. (2014), An et al. (2012), Al-Arabi and Khamis (1982), Popiel et al. (2007), Na and Chiou (1980), Chae and Chung (2011), Qiu et al. (2013), Chen and Hsu

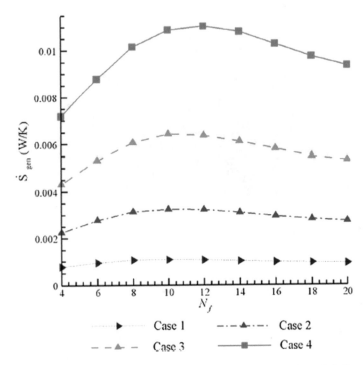

Fig. 2.15 Rate of entropy generation in cases 1–4 as a function of fin numbers (Sajedi et al. 2015)

(2007), Haldar et al. (2007), Mokheimer (2002), Elenbaas (1942) and Beckwith et al. (1990) have all worked with fins for natural convection heat exchanger applications.

Murali and Katte (2008) presented the performance of radiating pin fin having threads, grooves and taper on the outer surface. They concluded that the heat transfer rate from the radiator using threaded, grooved and tapered fin was about 1.2–3.7 times more than that from a solid radiating pin fin. Wilkins (1960), Kumar and Venkateshan (1994), Krishnaprakas (1996), Ramesh and Venkateshan (1997), Krikkis and Razelos (2002, 2003), Chung and Nguyen (1987), Schnurr et al. (1976), Black and Schoenhals (1968), Black (1973), Gorchakov and Panevin (1975, 1976), Bhise et al. (2002), Srinivasan and Katte (2004) and Holman (2000) have also worked on radiating fins.

In-line tube geometry is seldom used because it provides substantially lower performance than the staggered tube geometry. Effect of fin spacing is important (Fig. 2.18). Rich (1973, 1975) measured heat transfer and friction data for fin geometry. Equation (2.1) gives that the friction drag force which is the sum of the drag force on a bare tube bank and the drag caused by the fins (Rich 1973).

$$f_f = (\Delta p - \Delta p_t) \frac{2A_c \rho}{G^2 A_f} \tag{2.1}$$

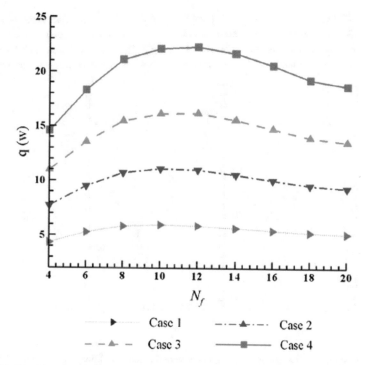

Fig. 2.16   Total heat loss in cases 1–4 as a function of fin numbers (Sajedi et al. 2015)

Both pressure drop contributions are evaluated at the same minimum area mass velocity. On many occasions, it may happen that Reynolds number based on hydraulic diameter will not correlate the effect of fin pitch.

Several investigators have observed conflicting behaviour of manifestation of flow: j factor may or may not have been affected by fin pitch, but also may or may not have been affected by row effect; Wang et al. (1996), Wang and Chi (2000), Yan and Sheen (2000), McQuiston (1978), Seshimo and Fujii (1991), Kayansayan (1993) and Abu Madi et al. (1998). Figures 2.19 and 2.20 show the j and f versus Re_{dh} and average heat transfer coefficients for plain plate-finned tubes, respectively.

McQuiston (1978), Gray and Webb (1986), Kim et al. (1999) and Wang et al. (2000) correlated j and f data versus Reynolds number for plain fins on staggered tube arrangements; the accuracy level of predictions, however, widely vary.

The Gray and Webb (1986) heat transfer correlation for four or more tube rows of staggered tube geometry is given by Eq. (2.2). The correlation for rows less than four needs a correction factor given by Eq. (2.3).

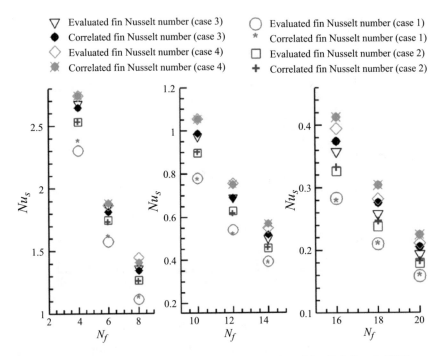

∇ Evaluated fin Nusselt number (case 3)   ○ Evaluated fin Nusselt number (case 1)
◆ Correlated fin Nusselt number (case 3)   * Correlated fin Nusselt number (case 1)
◇ Evaluated fin Nusselt number (case 4)   ☐ Evaluated fin Nusselt number (case 2)
✳ Correlated fin Nusselt number (case 4)   + Correlated fin Nusselt number (case 2)

Fig. 2.17 Variation of average Nusselt number with the number of fins (Sajedi et al. 2015)

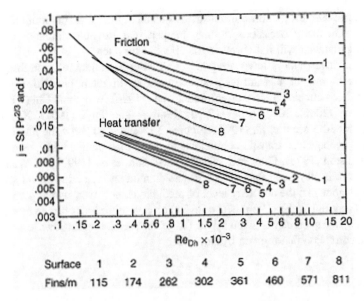

Surface	1	2	3	4	5	6	7	8
Fins/m	115	174	262	302	361	460	571	811

Fig. 2.18 Heat transfer and friction characteristics of a four-row plain plate fin heat exchanger for different fin spacing (Webb and Kim 2005)

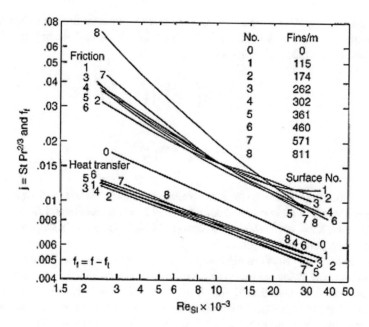

Fig. 2.19 Plot of the *j* factor and the fin friction vs. Re_{st} (Webb and Kim 2005)

Fig. 2.20 Average heat transfer coefficients for plain plate-finned tubes (571 fins/m) having one to six rows (Webb and Kim 2005)

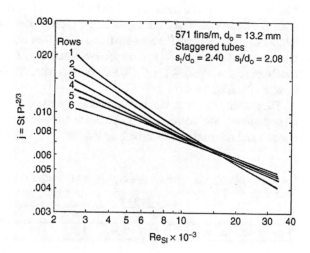

$$j_4 = 0.14 Re_d^{-0.328} \left(\frac{S_t}{S_l}\right)^{-0.502} \left(\frac{s}{d_0}\right)^{0.031} \tag{2.2}$$

$$\frac{j_N}{j_4} = 0.991 \left[2.24 Re_d^{-0.092} \left(\frac{N}{4}\right)^{-0.031}\right]^{0.607(4-N)} \tag{2.3}$$

McQuiston's (1978) correlation assumes that the pressure drop is composed of two terms; the first term is for the drag force on the fins and the second term for the drag force on the tubes (Eqs. 2.4 and 2.5).

$$f = f_f \frac{A_f}{A} + f_t \left(1 - \frac{A_f}{A}\right)\left(1 - \frac{t}{p_f}\right) \tag{2.4}$$

$$f_f = 0.508 Re_d^{-0.521} \left(\frac{S_t}{d_0}\right)^{1.318} \tag{2.5}$$

The friction factor with the tubes is obtained from a correlation for flow normal to a staggered bank of plain tubes. Zukauskas (1972) and Incropera and Dewitt (2001) give the tube bank correlation. McQuiston (1978) correlation based on the same data, however, does a poor job.

Mon and Gross (2004) numerically examined the fin-spacing effects by three-dimensional simulation of four-row tube bundles placed in staggered and in-line arrangements. It is complicated and difficult to understand geometrically complex bundles related to heat transfer characteristics. Thus, numerical model may help in better understanding and explanation. Saboya and Sparrow (1974, 1976), Sheu and Tsai (1999), Xi and Torikoshi (1996), Fiebig et al. (1995), Torikoshi (1994), Kaminski (2002), Kaminski and Groß (2003) and Romero-Méndez et al. (2000) studied finned tube heat exchangers and evaluated the influence of fin spacing but neither of them worked for annular-finned tube heat exchangers. They used the K-ε turbulence model and adopted respective equations. They simulated for Reynolds number range of $8600 \le Re \le 43,000$ and presented Table 2.3 for the dimension of bundle of testing tube.

They simulated both the in-line and staggered arrangement of tubes and concluded global flow behaviour, local flow behaviour, thermal boundary layer development and fin spacing effects on heat transfer and pressure drop characteristics. The

Table 2.3 Dimensions of tube bundles (Mon and Gross 2004)

	Staggered					In-line		
	s1	s2	s3	s4	s5	i1	i2	i3
Tube outside diameter, d	24	24	24	24	24	24	24	24
Fin diameter, d_f	34	34	34	44	44	34	34	34
Fin height, h_f	5	5	5	10	10	5	5	5
Fin thickness, t_f	0.5	0.5	0.5	0.5	0.5	0.5	0.5	0.5
Fin spacing, s	1.6	2	4	0.7	2	1.6	2	4
Fin pitch, $S_f = s + t_f$	2.1	2.5	4.5	1.2	1.2	2.1	2.5	4.5
Transverse tube pitch, S_t	40.8	40.8	40.8	52.8	52.8	40.8	40.8	40.8
Longitudinal tube pitch, S_1	35.33	35.33	35.33	45.73	45.73	40.8	40.8	40.8
Number of rows, n	4	4	4	4	4	4	4	4

All dimensions are in mm

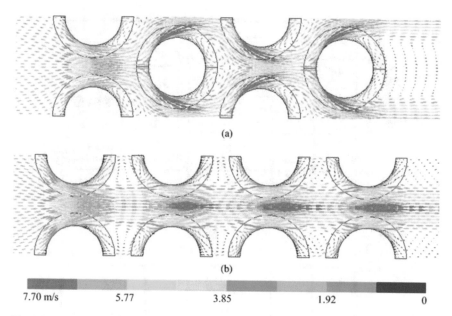

7.70 m/s   5.77   3.85   1.92   0

Fig. 2.21 Global velocity distributions for (**a**) staggered and (**b**) in-line arrangements at $Re = 8600$ (Mon and Gross 2004)

results of global velocity distribution are presented in Fig. 2.21, and it is found that main stream has been encountered by larger surface area in staggered array, whereas larger wake regions are in in-line array. Their simulated result shows that secondary vortex has developed. They plotted the results which included heat transfer coefficient versus fin spacing to height ratio in Fig. 2.22a and heat transfer coefficient versus pressure drop in Fig. 2.22b for both staggered and in-line bundles. The simulation presented that heat transfer coefficient increased to 19% in all cases with increased S/h_f from 0.32 to 0.8 simultaneously, whereas in case of staggered arrangement, the heat transfer coefficient was found to be increased up to $S/h_f = 0.32$ and remained constant for further increase in S/h_f ratio. They found that boundary layers between fins departing from each other for staggered arrangement. Pressure drop found to be decreased with both the arrangement as S/h_f increased.

Seshimo and Fujii (1991) gave a more generalized correlation for staggered banks of plain fins having one to five tube rows. They correlated one- and two-row data in terms of entrance length by Eqs. (2.6) and (2.7).

$$Nu = 2.1(X_{Dv})^n \tag{2.6}$$

$$fLD_v = c_1 + c_2(X_{Dv}) - m \tag{2.7}$$

For three or more rows, the entrance length-based correlations do not do justice to the data over the entire Reynolds number range ($200 < Re_{Dh} < 800$).

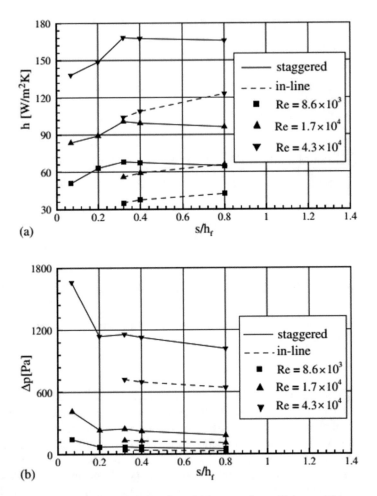

Fig. 2.22 Effects of fin spacing to height ratio on (**a**) heat transfer coefficient and (**b**) pressure drop for staggered and in-line bundles (Mon and Gross 2004)

Use of smaller diameter finned tube heat exchanger is the recent trend; Kim et al. (1999) improved Gray and Webb (1986) correlation by including the data of Wang and Chi (2000) and Youn (1997) for heat exchangers having smaller diameter tubes. The improvement in prediction was noteworthy. A more general correlation is that of Wang et al. (2000).

The Kim et al. (1999) correlation (for three or more tube rows) is given by a set of Eqs. (2.8), (2.9) and (2.10).

$$j_3 = 0.163 Re_d^{-0.369} \left(\frac{S_t}{S_l}\right)^{0.106} \left(\frac{s}{d_0}\right)^{0.0138} \left(\frac{S_t}{d_0}\right)^{0.13} \quad (N \geq 3) \qquad (2.8)$$

$$\frac{j_N}{j_3} = 1.043 \left[Re_d^{-0.14}\right] \cdot \left[\left(\frac{S_t}{S_l}\right)^{-0.564} \left(\frac{s}{d_0}\right)^{-0.123} \left(\frac{S_t}{d_0}\right)^{1.17}\right]^{(3-N)} \quad (N = 1, 2)$$
$$(2.9)$$

$$f_f = 1.455 Re_d^{-0.656} \left(\frac{S_t}{S_l}\right)^{-0.347} \left(\frac{s}{d_0}\right)^{-0.134} \left(\frac{S_t}{d_0}\right)^{1.23} \qquad (2.10)$$

and they used the Jacob (1938) correlation. Equation (2.4) is used to calculate the friction factor of the heat exchanger. In-line tube geometries are not good because tube bypass effects substantially degrade the performance of an in-line tube arrangement (Schmidt 1963).

References

Abu Madi M, Johns RA, Heikal MR (1998) Performance characteristics correlation for round tube and plate finned heat exchangers. Int J Refrig 21:507–517

Ahmadi M, Mostafavi G, Bahrami M (2014) Natural convection from rectangular interrupted fins. Int J Therm Sci 82(1):62–71

Alam I, Ghoshdastidar PS (2002) A study of heat transfer effectiveness of circular tubes with internal longitudinal fins having tapered lateral profiles. Int J Heat Mass Transf 45 (6):1371–1376

Al-Arabi M, Khamis M (1982) Natural convection heat transfer from inclined cylinders. Int J Heat Mass Transf 25(I):3–15

An BH, Kim HJ, Kim DK (2012) Nusselt number correlation for natural convection from vertical cylinders with vertically oriented plate fins. Exp Thermal Fluid Sci 41:59–66

Atayılmaz SO, Teke I (2009) Experimental and numerical study of the natural convection from a heated horizontal cylinder. Int Commun Heat Mass Transf 36:731–738

Atayılmaz SO, Teke I (2010) Experimental and numerical study of the natural convection from a heated horizontal cylinder wrapped with a layer of textile material. Int Commun Heat Mass 37:58–67

Atkinson KN, Drakulic R, Heikal MR, Cowell TA (1998) Two- and three-dimensional numerical models of flow and heat transfer over louvered fin arrays in compact heat exchangers. Int J Heat Mass Transf 41:4063–4080

Awad M, Muzychka YS (2011) Models for pressure drop and heat transfer in air cooled compact wavy fin heat exchangers. J Enhanc Heat Transf 18(3):191–207

Bahrami S, Rahimian MH, Mohammadbeigi H, Hosseinimanesh H (2012) Thermal-hydraulic study of multi-louvered fins in compact heat exchangers and recommendations for improvement. J Enhanc Heat Transf 19(1):53–61

Beckwith TG, Marangoni RD, Lienhard JH (1990) Mechanical measurements, 5th edn. Addison-Wesley Publishing Company, New York, pp 45–112

Bhise NV, Katte SS, Venkateshan SP (2002) A numerical study of corrugated structure for space radiators. In: 16th national and 5th ISHMT–ASME heat and mass transfer conference Kolkata, pp 520–526

Bilen K, Yapici S (2002) Heat transfer from a surface fitted with rectangular blocks at different orientation angle. Heat Mass Transf 38:649–655

Black WZ (1973) Optimization of the directional emission from V-groove and rectangular cavities. J Heat Transf 95:31–36

Black WZ, Schoenhals RJ (1968) A study of directional radiation properties of specially pre pared 'V'-groove cavities. J Heat Transf 90:420–428

Chae MS, Chung BJ (2011) Effect of pitch-to-diameter ratio on the natural convection heat transfer of two vertically aligned horizontal cylinders. Exp Thermal Fluid Sci 66:5321–5329

Chang YJ, Hsu KC (2000) Generalized friction correlation for louver fin geometry. Int J Heat Mass Transf 43(12):2237–2243

Chang Y, Wang C (1996) Air-side performance of brazed aluminium heat exchangers. J Enhanc Heat Transf 3(1):15–28

Chang Y, Wang C (1997) A generalized heat transfer correlation for louvered fin geometry. Int Heat Transf 40(3):533–544

Chang Y, Wang C, Chang W (1994) Heat transfer and flow characteristics of automotive brazed aluminium heat exchangers. ASHRAE Trans 100(2):643–652

Chen HT, Hsu WL (2007) Estimation of heat transfer coefficient on the fin of annular-finned tube heat exchangers in natural convection for various fin spacings. Int J Heat Mass Transf 50:1750–1761

Chung BTF, Nguyen LD (1987) Thermal analysis and optimum design for radiating spine of various geometries. In: Proceedings of the international symposium on heat transfer science and technology Beijing People's Republic of China October 15–18 198 (A87-33101 13-34). Hemisphere Publishing Corp, Washington, DC, pp 510–517

Dillen EL, Webb RL (1994) Rationally based heat transfer and friction correlations for the louver fin geometry. SAE Tech Paper Ser 94050:600–607

Dong J, Chen J, Chen Z, Zhang W, Zhou Y (2007) Heat transfer and pressure drop correlations for the multi-louvered fin compact heat exchangers. Energy Convers Manag 48:1506–1515

Eckert E, Irvine T (1960) Pressure drop and heat transfer in a duct with triangular cross-section. ASME J Heat Transf 83:125–136

Elenbaas W (1942) Heat dissipation of parallel plates by free convection. Physica 9(1):1–28

El-Saed SA, Mohamed SM, Abdel-Latif AM, Abouda AE (2002) Investigation of turbulent heat transfer and fluid flow in longitudinal rectangular fin-arrays of different geometries and shrouded fin array. Exp Thermal Fluid Sci 26:879–900

Fabbri G (1998) Heat transfer optimization in internally finned tubes under laminar flow conditions. Int J Heat Mass Transf 41(10):1243–1253

Fabbri G (1999) Optimum profiles for asymmetrical longitudinal fins in cylindrical ducts. Int J Heat Mass Transf 4(23):511–523

Fiebig M, Gorsse-Georgemann A, Chen Y, Mitra NK (1995) Conjugate heat transfer of a finned tube part A: heat transfer behaviour and occurrence of heat transfer reversal. Numer Heat Transf Part A 28:133–146

Gorchakov VS, Panevin IG (1975) Effectiveness of radiating fins covered with V-shaped grooves. http://techreports.iarc.nasa.gov/egiin/ NTRS

Gorchakov VS, Panevin IG (1976) Efficiency of radiating fins covered with V-shaped grooves. J High Temp 13(4):733–738

Gray DL, Webb RL (1986) Heat transfer and friction correlations for plate fin-and-tube heat exchangers having plain fins. In: Heat transfer 1986. Proceedings of the eighth international heat transfer conference, pp 2745–2750

Haghighi SS, Goshayeshi HR, Safaei MR (2018) Natural convection heat transfer enhancement in new designs of plate-fin based heat sinks. Int J Heat Mass Transf 125:640–647

Haldar SC, Kochhar GS, Manohar K, Sahoo RK (2007) Numerical study of laminar free convection about a horizontal cylinder with longitudinal fins of finite thickness. Int J Therm Sci 46:692–698

Holman JP (2000) Experimental methods for engineering. Ch. 2 and 3. McGraw-Hill, New York

Incropera FP, DeWitt DP (2001) Fundamentals of heat mass transfer, 5th edn. Wiley, New York

Islam MD, Oyakawa K, Yaga M (2008) Heat transfer enhancement from a surface affixed with rectangular fins of different patterns and arrangement in duct flow. J Enhanc Heat Transf 15 (1):31–50

Jacob ML (1938) Heat transfer and flow resistance in cross flow of gases over tube banks. Trans ASME 60:384

Jeon D, Byon C (2017) Thermal performance of plate fin heat sinks with dual-height fins subject to natural convection. Int J Heat Mass Transf 113:1086–1092

Ji C, Qin Z, Low Z, Dubey S, Choo FH, Duan F (2018) Non-uniform heat transfer suppression to enhance PCM melting by angled fins. Appl Therm Eng 129:269–279

Joo Y, Kim SJ (2015) Comparison of thermal performance between plate-fin and pin fin heat sinks in natural convection. Int J Heat Mass Transf 83:345–356

Junqi D, Jiangping C, Zhijiu C, Yimin Z, Wenfeng Z (2007) Heat transfer and pressure drop correlations for the wavy fin and flat tube heat exchangers. Appl Therm Eng 27 (11–12):2066–2073

Kadle DS, Sparrow EM (1986) Numerical and experimental study of turbulent heat transfer and fluid flow in longitudinal fin array. ASME J Heat Transf 108:16–23

Kaminski S (2002) Numerische Simulation der luftseitigen Stromungs-und Warmetransportvorgange in Lamellenrohr-Warmeubertragern. Techn. Univ. Bergakad, Freiberg

Kaminski S, Groß U (2003) Luftseitige Transportprozesse in Lamellenrohrbundeln—numerische Untersuchung. Ki Luft und Kaltetechnik (5):220–224

Kayansayan N (1993) Heat transfer characterization of flat plain fins and round tube heat exchangers. Exp Thermal Fluid Sci 6(3):263–272

Kim N-H, Youn B, Webb RL (1999) Air-side heat transfer and friction correlations for plain fin-and-tube heat exchangers with staggered tube arrangements. J Heat Transf 121(3):662–667

Krikkis RN, Razelos P (2002) Optimum design of spacecraft radiators with longitudinal rectangular and triangular fins. J Heat Transf 124:805–811

Krikkis RN, Razelos P (2003) The optimum design of radiating and convective-radiating circular fins. J Heat Transf Eng 24(3):17–41

Krishnaprakas CK (1996) Optimum design of radiating rectangular plate fin array extending from a plane wall. J Heat Transf 118:490–493

Kumar SS, Venkateshan SP (1994) Optimized tubular radiator with annular fins on a non-isothermal base. Int J Heat Fluid Flow 15:399–409

Kumar R (1997) Three-dimensional natural convective flow in a vertical annulus with longitudinal fins. Int J Heat Mass Transf 40(14):3323–3334

Lawson MJ, Thole KA (2008) Heat transfer augmentation along the tube wall of a louvered fin heat exchanger using practical delta winglets. Int J Heat Mass Transf 51:2346–2360

Lee M, Kim HJ, Kim DK (2016) Nusselt number correlation for natural convection from vertical cylinders with triangular fins. Appl Therm Eng 93:1238–1247

Lin Y-T, Hwang Y-M, Wang C-C (2002) Performance of the herringbone wavy fin under dehumidifying conditions. Int J Heat Mass Transf 45:5035–5044

Liu XY, Jensen MK (1999) Numerical investigation of turbulent flow and heat transfer in internally finned tubes. J Enhanc Heat Transf 6(2–4):105–119

Lyman AC, Stephan RA, Thole KA, Zhang LW, Memory SB (2002) Scaling of heat transfer coefficients along louvered fins. Exp Thermal Fluid Sci 26:547–563

McQuiston FC (1978) Correlation of heat, mass, and momentum transport coefficients for plate-fin-tube heat transfer for surfaces with staggered tube. ASHRAE Trans 54(Part 1):294–309

Micheli L, Reddy K, Mallick TK (2016) Experimental comparison of micro-scaled plate-fins and pin-fins under natural convection. Int Commun Heat Mass Transf 75:59–66

Mohammadian SK, Zhang Y (2017) Cumulative effects of using pin fin heat sink and porous metal foam on thermal management of lithium-ion batteries. Appl Therm Eng 118:375–384

Mokheimer EMA (2002) Performance of annular fins with different profiles subject to variable heat transfer coefficient. Int J Heat Mass Transf 45:3631–3642

Molki M, Faghri M, Ozbay O (1995) A correlation for heat transfer and wake effect in the entrance region of an inline array of rectangular blocks simulating electronic components. ASME J Heat Transf 117:40–46

Mon MS, Gross U (2004) Numerical study of fin-spacing effects in annular-finned tube heat exchangers. Int J Heat Mass Transf 47(8–9):1953–1964

Muley A, Borghese J, Manglik RM, Kundu J (2002) Experimental and numerical investigation of thermal-hydraulic characteristics of wavy-channel compact heat exchanger. In: Proc. 12th international heat transfer conference France, vol 4, pp 417–422

Muley A, Borghese JB, White SL, Manglik RM (2006) Enhanced thermal-hydraulic performance of a wavy-plate fin compact heat exchanger: effect of corrugation severity. In: Proc. 2006 ASME international mechanical engineering congress and exposition (IMECE2006), Chicago, IL, USA, IMECE2006-14755

Murali JG, Katte SS (2008) Experimental investigation of threaded, grooved, and tapered radiating pin-fin. J Enhanc Heat Transf 15(3):199–209

Muzychka YS (1999) Analytical and experimental study of fluid friction and heat transfer in low Reynolds number flow heat exchangers. Ph. D Thesis. University of Waterloo, Waterloo, ON

Muzychka YS, Kenway G (2009) A model for the thermal hydraulic characteristics of the offset strip fin array for large Prandtl number liquids. J Enhanc Heat Transf 16(1):73–92

Na TY, Chiou JP (1980) Turbulent natural convection over a slender circular cylinder. Warme Stoffubertrag 14:157–164

Ohara J, Koyama S (2012) Falling film evaporation of pure refrigerant HCFC123 in a plate-fin heat exchanger. J Enhanc Heat Transf 19(4):301–311

Olson DA (1992) Heat transfer in thin, compact heat exchangers with circular, rectangular, or pin-fin flow passages. ASME J Heat Transf 114:373–382

Oyakawa K, Furukawa Y, Taira T, Senaha I (1993) Effect of vortex generators on heat transfer enhancement in a duct. In: Proceedings of the experimental heat transfer, fluid mechanics and thermodynamics Honolulu, Hawaii, vol 1, pp 633–640

Park J, Ligrani PM (2005) Numerical predictions of heat transfer and fluid flow characteristics for seven different dimpled surfaces in a channel. Numer Heat Transf Part A Appl 47(3):209–232

Park KT, Kim HJ, Kim DK (2014) Experimental study of natural convection from vertical cylinders with branched fins. Exp Thermal Fluid Sci 54:29–37

Perrotin T, Clodie D (2004) Thermal-hydraulic CFD study in louvered fin-and-flat-tube heat exchangers. Int J Refriger 27:422–432

Popiel CO, Wojtkowiak J, Bober K (2007) Laminar free convective heat transfer from isothermal vertical slender cylinder. Exp Thermal Fluid Sci 32(2007):607–613

Qiu Y, Tian M, Guo Z (2013) Natural convection and radiation heat transfer of an externally-finned tube vertically placed in a chamber. Heat Mass Transf 49:405–412

Ramesh N, Venkateshan SP (1997) Optimum finned tubular space radiator. Heat Transf Eng 18:69–87

Rich DG (1973) The effects of fin spacing on the heat transfer and friction performance of multi-row, smooth plate fin-and-tube heat exchangers. ASHRAE Trans 79(Part 2):137–145

Rich DG (1975) Effect of the number of tube rows on heat transfer performance of smooth plate fin-and-tube heat exchangers. ASHRAE Trans 81(Part 1):307–319

Romero-Méndez R, Sen M, Yang KT, McClain R (2000) Effect of fin spacing on convection in a plate fin and tube heat exchanger. Int J Heat Mass Transf 43(1):39–51

Rush TA, Newell TA, Jacobi AM (1999) An experimental study of flow and heat transfer in sinusoidal wavy passages. Int J Heat Mass Transf 42(9):1541–1553

Saad AE, Sayed AE, Mohamed EA, Mohamed MS (1997) Experimental study of turbulent flow inside a circular tube with longitudinal interrupted fins in the streamwise direction. Exp Thermal Fluid Sci 15(1):1–15

Saboya FEM, Sparrow EM (1974) Local and average transfer coefficients for one-row plate fin and tube heat exchanger configurations. J Heat Transf 96(3):265–272

Saboya FEM, Sparrow EM (1976) Transfer characteristics of two-row plate fin and tube heat exchanger configurations. Int J Heat Mass Transf 19(1):41–49

Sajedi R, Taghilou M, Jafari M (2015) Experimental and numerical study on the optimal fin numbering in an external extended finned tube heat exchanger. Appl Therm Eng 83:139–146

Sarkhi AA, Nada EA (2005) Characteristics of forced convection heat transfer in vertical internally finned tube. Int Commun Heat Mass Transf 32:557–564

Schmidt TE (1963) Der Warmeiibergang an Rippenrohre and die Berechnung von Rohrbundel-Warmeaustauschern, Kaltetechnik, Band 15, Heft 12

Schnurr NM, Townsend MA, Shapiro AB (1976) Optimization of radiating fin arrays with respect to weight. ASME Trans J Heat Transf 98:643–648

Seshimo Y, Fujii M (1991) An experimental study on the performance of plate fin and tube heat exchangers at low Reynolds numbers. In: Proceedings of the ASME-JSME thermal engineering joint conference, vol 4, pp 449–454

Sheik Ismail L, Ranganayakulu C, Shah RK (2009) Numerical study of flow patterns of compact plate-fin heat exchangers and generation of design data for offset and wavy fins. Int J Heat Mass Transf 52(17–18):3972–3983

Sheik Ismail L, Velraj R, Ranganayakulu C (2010) Studies on pumping power in terms of pressure drop and heat transfer characteristics of compact plate-fin heat exchangers—a review. Renew Sust Energ Rev 14(1):478–485

Sheu TW, Tsai SF (1999) A comparison study on fin surfaces in finned-tube heat exchangers. Int J Numer Methods Heat Fluid Flow 9(1):92–106

Shih TH, Liou WW, Shabbrir A, Yang ZG, Zhu J (1995) A new k–e eddy viscosity model for high Reynolds number turbulent flows. Comput Fluids 24(3):227–238

Sparrow EM, Niethammer JE, Chaboki A (1982) Heat transfer and pressure drop characteristics of arrays of rectangular modules encountered in electronic equipment. Int J Heat Mass Transf 25:961–973

Sparrow EM, Vemuri SB, Kadle D (1983) Enhanced and local heat transfer, pressure drop, and flow visualization for arrays of block-like electronic components. Int J Heat Mass Transf 26:689–699

Srinivasan K, Katte SS (2004) Analysis of grooved space radiator. In: Proceedings of the 17th national and 6th ISHMT-ASME heat and mass transfer confence, Kalpakkam, vol 12

Sunden B, Svantesson J (1992) Correlation of j-and f-factors for multilouvered heat transfer surfaces. In: Proc. 3rd UK national conference on heat transfer, pp 805–811

Taghilou M, Ghadimi B, Seyyedvalilu MH (2014) Optimization of double pipe fin pin heat exchanger using entropy generation minimization. IJE Trans C Aspects 27(9):1445–1454

Torii K, Yanagihara JI (1997) A review on heat transfer enhancement by longitudinal vortices. J HTSJ 36(142):73–86

Torikoshi K (1994) Flow and heat transfer performance of a plate-fin and tube heat exchanger. Heat Transf 4:411–416

Turk AY, Junkhan GH (1986) Heat transfer enhancement downstream of vortex generators on a flat plate. In: Tien CL, Carey VP, Ferrell JK (eds) Heat transfer, vol 6. Hemisphere, Washington, pp 2903–2908

Wang C-C, Chi K-Y (2000) Heat transfer and friction characteristics of plain fin-and-tube heat exchangers, part I: new experimental data. Int J Heat Mass Transf 43:2681–2691

Wang C-C, Chen P-Y, Jang J-Y (1996) Heat transfer and friction characteristics of convex-louver fin-and-tube heat exchangers. Exp Heat Transf 9:61–78

Wang C-C, Chi K-Y, Chang C-J (2000) Heat transfer and friction characteristics of plain fin-and-tube heat exchangers, part II: correlation. Int J Heat Mass Transf 43:2693–2700

Wang QW, Lin M, Zeng M (2008a) Effect of blocked core-tube diameter on heat transfer performance of internally finned tubes. Heat Transf Eng 29(1):107–115

Wang QW, Lin M, Zeng M, Tian L (2008b) Computational analysis of heat transfer and pressure drop performance for internally finned tubes with three different longitudinal wavy fins. Heat Mass Transf 45:147–156

Wang QW, Lin M, Zeng M, Tian L (2008c) Investigation of turbulent flow and heat transfer in periodic wavy channel of internally finned tube with blocked core tube. ASME J Heat Transf 130(6). Article No.: 061801

Webb RL, Kim NY (2005) Principles of enhanced heat transfer. Taylor & Francis, New York

Wilkins JE Jr (1960) Minimizing the mass of thin radiating fins. J Aerospace Sci 27:145–146

Xi GN, Torikoshi K (1996) Computation and visualizationof flow and heat transfer in finned tube heat exchangers. In: International symposium on heat transfer, Tsinhua University, Beijing China (7.10–11.10), pp 632–637

Yan W-M, Sheen P-J (2000) Heat transfer and friction characteristics of finand-tube heat exchangers. Int J Heat Mass Transf 43:1651–1659

Yang YT, Peng HS (2009) Investigation of planted pin fins for heat transfer enhancement in plate fin heat sink. Microelectron Reliab 49(2):163–169

Yang Y, Li Y, Si B, Zheng J (2017) Heat transfer performances of cryogenic fluids in offset strip fin-channels considering the effect of fin efficiency. Int J Heat Mass Transf 114:1114–1125

Yazicioğlu B, Yüncü H (2007) Optimum fin spacing of rectangular fins on a vertical base in free convection heat transfer. Heat Mass Transf 44(1):11–21

Youn B (1997) Internal report. Samsung Electric Corp

Yu B, Tao WQ (2004) Pressure drop and heat transfer characteristics of turbulent flow in annular tubes with internal wave-like longitudinal fins. Heat Mass Transf 40:643–651

Yu B, Nie JH, Wang QW, Tao WQ (1999) Experimental study on the pressure drop and heat transfer characteristics of tubes with internal wave-like longitudinal fins. Heat Mass Transf 35:65–73

Yu X, Feng J, Feng Q, Wang Q (2005) Development of a plate-pin fin heat sink and its performance comparisons with a plate fin heat sink. Appl Therm Eng 25(2):173–182

Zaretabar M, Asadian H, Ganji D (2018) Numerical simulation of heat sink cooling in the mainboard chip of a computer with temperature dependent thermal conductivity. Appl Therm Eng 130:1450–1459

Zeitoun O, Hegazy AS (2004) Heat transfer for laminar flow in internally finned pipes with different fin heights and uniform wall temperature. Heat Mass Transf 40:253–259

Zhang J (2005) Numerical simulations of steady low-Reynolds-number flows and enhanced heat transfer in wavy plate-fin passages. Ph.D. thesis, University of Cincinnati

Zhang J, Muley A, Borghess JB, Manglik RM (2003) Computational and experimental study of enhanced laminar flow heat transfer in three dimensional sinusoidal wavy-plate-fin channels. In: Proceedings of the 2003 ASME summer heat transfer conference, Nevada, USA, HT2003-47148

Zhang J, Kundu J, Manglik RM (2004) Effect of fin waviness and spacing on the lateral vortex structure and laminar heat transfer in wavy-plate-fin cores. Int J Heat Mass Trans 47:1719–1730

Zukauskas A (1972) Heat transfer from tubes in crossflow. In: Hartnett JP, Irvine TF (eds) Advances in heat transfer, vol 8. Academic Press, New York, pp 93–160

Chapter 3
Circular Fins with Staggered Tubes, Low Integral Fin Tubes

Circular fins with staggered tubes are frequently used in the process industries and in combination with heat-recovery equipment. These are extruded fins or helically wrapped fins on circular tubes. There may be both plain and enhanced fin geometries. For high fins ($e/d_o > 0.2$), a staggered tube layout is used. Good amount of performance data and several heat transfer and pressure drop correlations are available. Staggered tube arrangement has been made for six or more tube rows deep. Tube bank variables like d_o, S_p and S_l and fin geometry variables like t, e and s and the number of tube rows are carefully considered for the development of correlations; Webb (1987) provided a good review of this regarding published data and correlations.

Jayavel and Tiwari (2010) investigated the effect of longitudinal and transverse tube spacing on the performance of staggered tube bundle fitted in the fin-tube heat exchanger on the flow and heat transfer characteristics. They arranged the tube bundles in staggered fashion and subjected to cross flow of air. They conducted three-dimensional numerical simulation by using finite volume computational code. They used multiple pairs of vortex generator (VG) and analysed the thermal hydraulic properties in steady laminar flow region. Figure 3.1 shows the schematic diagram of fin-tube heat exchanger. Table 3.1 shows the effect of vortex generators on the heat transfer, friction factor and performance index of heat exchanger. Table 3.2 presents the comparison of average Nusselt number, friction factors and performance index with longitudinal spacing of the tubes. It was observed that all the thermophysical parameters mentioned in Table 3.2 decreased with increase in longitudinal spacing of the tubes.

Figure 3.2 shows variation of average Nusselt numbers and average pressure drop with increasing values of longitudinal spacing of the tubes. Table 3.3 shows the effect of the transverse spacing of the Nusselt number, friction factors and performance index of the tubes. It is clear from the Table 3.3 that there is a gradual increase in the performance index as the transverse spacing of the tubes increase. They optimized the longitudinal and transverse tube spacing of the staggered tube bundle

© The Author(s), under exclusive license to Springer Nature Switzerland AG 2020
S. K. Saha et al., *Heat Transfer Enhancement in Externally Finned Tubes and Internally Finned Tubes and Annuli*, SpringerBriefs in Applied Sciences and Technology, https://doi.org/10.1007/978-3-030-20748-9_3

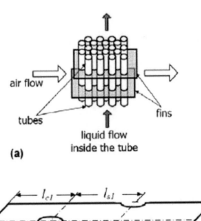

(a)

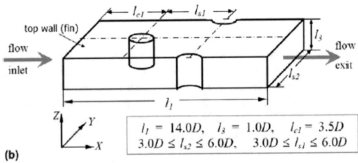

$$l_1 = 14.0D, \quad l_3 = 1.0D, \quad l_{c1} = 3.5D$$
$$3.0D \le l_{s2} \le 6.0D, \quad 3.0D \le l_{s1} \le 6.0D$$

(b)

Fig. 3.1 (**a**) Schematic of the fin-tube heat exchanger, (**b**) representation of the computational domain with dimensions (Jayavel and Tiwari 2010)

Table 3.1 Effect of vortex generators (Jayavel and Tiwari 2010)

Sl. no.	VG1	VG2	$\overline{Nu}_o$	f	η
1	No	No	12.4650	0.1631	1.00
2	Yes	No	14.8494	0.1861	1.14
3	No	Yes	14.2247	0.1742	1.12
4	Yes	Yes	16.3318	0.1981	1.23

$l_{s1} = l_{s2} = 4D$ and $Re_D = 400$

Table 3.2 Effect of the longitudinal spacing of the tubes (Jayavel and Tiwari 2010)

Sl. no.	l_{s1}	$\overline{Nu}_o$	f	η
1	3D	18.1558	0.2193	1.07
2	4D	16.3318	0.1981	1.00
3	5D	15.0958	0.1701	0.97
4	6D	14.0038	0.1697	0.90

$l_{s2} = 4D$, $Re_D = 400$ and in the presence of VG1 and VG2

in the presence of vortex generators such that to maximize the heat transfer enhancement and minimize the pressure drop. Figures 3.3 and 3.4 present the streamline plots and temperature distribution, respectively, near the bottom wall of the channel. Baker (1991), Buyruk et al. (1998), Fiebig (1995), Jahromi Bastani et al. (1999), Jain

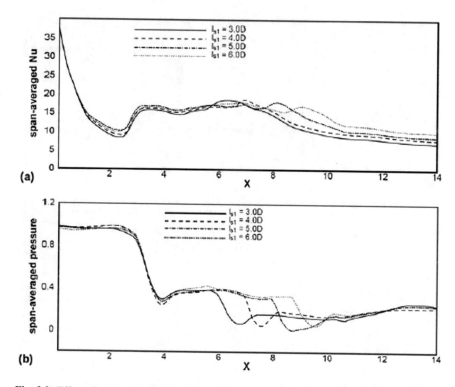

Fig. 3.2 Effect of the longitudinal spacing of tubes: (**a**) span-averaged Nusselt number variation near the bottom wall, (**b**) height-averaged pressure variation (Jayavel and Tiwari 2010)

Table 3.3 Effect of the transverse spacing of the tubes (Jayavel and Tiwari 2010)	Sl. no.	l_{s2}	$\overline{Nu}_o$	f	η
	1	$3D$	15.7044	0.1827	0.99
	2	$4D$	16.3318	0.1981	1.00
	3	$5D$	17.2063	0.2113	1.03
	4	$6D$	17.7103	0.2261	1.04

$l_{s1} = 4D$, $Re_D = 400$ and in the presence of VG1 and VG2

et al. (2003), Joardar and Jacobi (2008) and Kwak et al. (2005) investigated the effect of vortex generator in fin-tube heat exchanger.

Empirically based on Briggs and Young (1963) correlation for heat transfer and Robinson and Briggs (1966) correlation for isothermal pressure drop, Eqs. (3.1) and (3.2) are strongly recommended for four or more tube rows.

$$j = 0.134 Re_d^{-0.319} \left(\frac{s}{e}\right)^{0.2} \left(\frac{s}{t}\right)^{0.11} \tag{3.1}$$

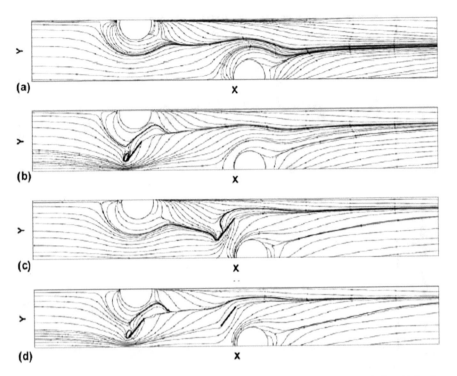

Fig. 3.3 Streamline plot near the bottom wall for different configurations of VGs: (a) in the absence of VG; (b) in the presence of VG1 only; (c) in the presence of VG2 only; (d) in the presence of both VG1 and VG2 (Jayavel and Tiwari 2010)

$$f_{tb} = 9.47 Re_d^{-0.316} \left(\frac{S_t}{d_0}\right)^{-0.927} \left(\frac{S_t}{S_d}\right)^{0.515} \tag{3.2}$$

Gianolio and Cuti (1981) data for 17 tube bank geometries containing one to six rows have been compared with Briggs and Young (1963) and Robinson and Briggs (1966) correlations. Robinson and Briggs (1966) correlation does not predict the data of Gianollo and Cuti (1981).

Rabas et al. (1981) has given a better j and f correlations for low fin heights and small fin spacing (Eqs. 6.16 and 6.17). Rabas and Taborek (1987) have surveyed correlations, row correction factors for low integral fin tube banks. Groehn (1977) and ESDU (1985) give other correlations. Rabas and Huber (1989) observed reduction of the j-factor with increased number of tube rows. Some information on in-line finned tube banks may be obtained from Schmidt (1963). Rabas et al. (1981) and Brauer (1964) correlation may be used for a staggered layout of low integral fins.

Kawaguchi et al. (2005) studied the heat transfer enhancement in forced convection phenomenon using finned tube banks. The serrated and spiral fin geometry has been considered, and their results have been compared. The spiral fin and serrated fin

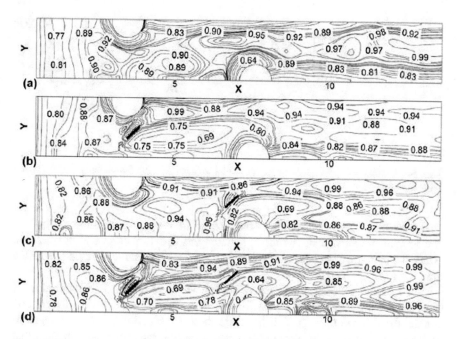

Fig. 3.4 Temperature distribution near the bottom wall for different configurations of VGs: (**a**) in the absence of VG; (**b**) in the presence of VG1 only; (**c**) in the presence of VG2 only; (**d**) in the presence of both VG1 and VG2 (Jayavel and Tiwari 2010)

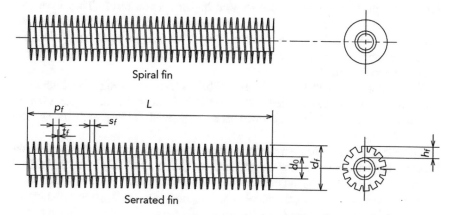

Fig. 3.5 Spiral fin and serrated fin geometry (Kawaguchi et al. 2005)

have been shown in Fig. 3.5. They observed superior performance of serrated fins over that of spiral fins, irrespective of the arrangement of the fins. The heat transfer coefficient for serrated fins was reported to be 1.25 times that of spiral fins, for smaller surface area of serrated fins. The increase in friction factor was observed to be greater for spiral fins as compared to that of serrated fins. They have explained

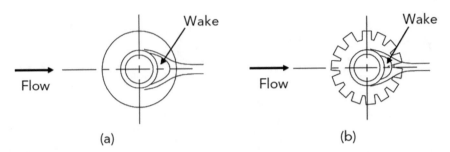

Fig. 3.6 Comparison of wake area in spiral and serrated fin. (**a**) Spiral fin. (**b**) Serrated fin (Kawaguchi et al. 2005)

that though the area of projection of the serrated fin is same as the area of spiral fin, the equivalent diameter in volume of serrated fins is only 96% of that of spiral fin. Thus, the pressure drop in case of spiral fin is greater than that in the case of serrated fin.

The friction at the fin area is generally expected to be high. But, it is easy for the flow around fins to be turned into wake area of the tube, thus decreasing the friction around the serrated fin area. This has been shown in Fig. 3.6. The rate of augmentation of heat transfer for serrated fins has been observed to be greater than that corresponding to spiral fins. The further increase in this rate of augmentation was observed for fins with larger pitch rather than those with smaller pitch. The friction factor was reported to be higher for serrated fins in case of large pitch values while for smaller pitch values spiral fins showed higher friction factor. They have also noted that arrangement of tubes in the tube banks had negligible effect on the heat transfer and pressure drop characteristics. They have also proposed correlations for Nusselt number and friction factor which agreed well with the experimental data with accuracy within 5%.

Kawamura et al. (1991a, b), Brauer (1961), Weyrauch (1969) and Yudin and Tokhtarowa (1964) have worked with spiral fins. Serrated fins have been used by Weierman et al. (1974), Ackerman and Brunsvold (1970) and Rabas and Eckels (1975).

The effect of fin angles, emissivity of fin surfaces and tube wall temperature on heat transfer enhancement was investigated by Qiu et al. (2013) in a longitudinal externally finned tube. The longitudinal fin was placed vertically in a small chamber. Three-dimensional numerical model was prepared for finding the solution of governing equation and boundary conditions and compared with the existing experimental results. Figure 3.6(a) shows the schematic diagram of a longitudinal externally finned tube. It was observed from the simulation results that mean Nusselt number increased with increase in Rayleigh number. It was observed that maximum radiative heat transfer was obtained at fin angle of 40° for the emissivity of 0.9 and maximum convective heat transfer occurred at fin angle of 45°.

Ratio of radiation heat transfer gradually decreased with increase in tube wall temperature. The highest ratio of radiation occurred at emissivity of 0.9 and tube wall temperature of 50°. The maximum heat transfer per unit mass occurred at fin angle 55° with fin surface emissivity 0.9. They found that total heat transfer rate varied and was almost directly linear proportional to the fin surface emissivity. It was also observed that convective heat transfer rate was greater than radiative heat transfer rate in the temperature range. Yu et al. (1999), Yu and Tao (2004), Wang et al. (2008, 2009), Fabbri (1998), Krupiczka et al. (2003), Wu and Tao (2007), Sun et al. (2002), Ouzzane and Galanis (2001) and Lozza and Merlo (2001) studied the optimization of the heat transfer with internally or externally or extruded finned tube.

References

Ackerman JW, Brunsvold AR (1970) Heat transfer and draft loss performance of extended surface tube banks. J Heat Transf Trans ASME 92:215–220

Baker CJ (1991) Oscillations of horseshoe vortex systems. ASME J Fluids Eng 113(3):489–495

Brauer H (1961) Spiral fin tubes for cross flow heat exchangers. Kaltetechnik 13:274–279

Brauer H (1964) Compact heat exchangers. Chem Prog Eng (London) 45(8):451–460

Briggs DE, Young EH (1963) Convection heat transfer and pressure drop of air flowing across triangular pitch banks of finned tubes. Chem Eng Prog Symp Ser 59(41):1–10

Buyruk E, Johnson MW, Owen I (1998) Numerical and experimental study of flow and heat transfer around a tube in cross-flow at low Reynolds number. Int J Heat Mass Transf 19:223–232

ESDU (1985) Low-fin staggered tube banks: heat transfer and pressure loss for turbulent single-phase crossflow. Eng Sci. Data Unit, ESDU Item No. 84016

Fabbri G (1998) Heat transfer optimization in internally finned tubes under laminar flow conditions. Int J Heat Mass Transf 41(10):1243–1253

Fiebig M (1995) Vortex generators for compact heat exchangers. J Enhanc Heat Transf 2:43–61

Gianollo E, Cuti F (1981) Heat transfer coefficients and pressure drop for air coolers with different numbers of rows under induced and forced draft. Heat Transf Eng 3(1):38–48

Groehn HG (1977) Flow and heat transfer studies of a staggered tube bank heat exchanger with low fins at high Reynolds numbers. Central Library of the Julich Nuclear Research Center GmbH, Julich, Germany

Jahromi Bastani AA, Mitra NK, Biswas G (1999) Numerical investigations on enhancement of heat transfer in a compact fin-and-tube heat exchanger using delta winglet type vortex generators. J Enhanc Heat Transf 6(1):1–11

Jain A, Biswas G, Maurya D (2003) Winglet-type vortex generators with common-flow-up configuration for fin-tube heat exchangers. Numer Heat Transf A 43(2):201–219

Jayavel S, Tiwari S (2010) Effect of tube spacing on heat transfer performance of staggered tube bundles in the presence of vortex generators. J Enhanc Heat Transf 17(3):271–291

Joardar A, Jacobi AM (2008) Heat transfer enhancement by winglet-type-vortex generator arrays in compact plain-fin-and-tube heat exchangers. Int J Refrig 31(1):87–97

Kawaguchi K, Okui K, Kashi T (2005) Heat transfer and pressure drop characteristics of finned tube banks in forced convection (comparison of heat transfer and pressure drop characteristics of serrated and spiral fins). J Enhanc Heat Transf 12(1):1–20

Kawamura T et al (1991a) Study on heat transfer and pressure drop characteristics of finned tube banks (1st report heat transfer characteristics and predicting equations of circular finned tube banks). Trans Jpn Soc Mech Eng B 57(537):1752–1758

Kawamura T et al (1991b) Study on heat transfer and pressure drop characteristics of finned tube banks (2nd report pressure drop characteristics and predicting equations of circular finned tube banks). Trans Jpn Soc Mech Eng B 57(537):1759–1764

Krupiczka R, Rotegel A, Walczyk H, Dobner L (2003) An experimental study of convective heat transfer from extruded type helical finned tubes. Chem Eng Process 42(1):29–38

Kwak KM, Torii K, Nishino K (2005) Simultaneous heat transfer enhancement and pressure loss reduction for finned-tube bundles with the first or two transverse rows of built-in winglets. Exp Thermal Fluid Sci 29:625–632

Lozza G, Merlo U (2001) An experimental investigation of heat transfer and friction losses of interrupted and wavy fins for fin-and-tube heat exchangers. Int J Refrig 24:409–416

Ouzzane M, Galanis N (2001) Numerical analysis of mixed convection in inclined tubes with external longitudinal fins. Sol Energy 71(3):199–211

Qiu Y, Tian M, Guo Z (2013) Natural convection and radiation heat transfer of an externally-finned tube vertically placed in a chamber. Heat Mass Transf 49(3):405–412

Rabas TI, Huber FV (1989) Row number effects on the heat transfer performance of in-line finned tube banks. Heat Trans Eng 10(4):19–29

Rabas TJ, Eckels PW (1975) Heat transfer and pressure drop performance of segmented extended surface tube bundles. In: Proc. AIChE-ASME heat transfer conference, San Francisco, pp 1–8

Rabas TJ, Eckels PW, Sabatino RA (1981) The effect of fin density on the heat transfer and pressure drop performance of low finned tube banks. Chem Eng Commun 10(1):127–147

Rabas TJ, Taborek J (1987) Survey of turbulent forced-convection heat transfer and pressure drop characteristics of low-finned tube banks in cross flow. Heat Trans Eng 8(2):49–62

Robinson KK, Briggs DE (1966) Pressure drop of air flowing across triangular pitch banks of finned tubes. Chem Eng Prog Symp Ser 62(64):177–184

Schmidt E (1963) Heat transfer at finned tubes and computations of tube bank heat exchangers. Kaltetechnik 15(4): 98 and 15(12): 370

Sun FZ, Zhang MY, Huang XY, Shi YT, Dong XG (2002) Experimental study of enhanced heat transfer with Ni-based impacted longitudinal finned tubes. J Hydrodyn 17(4):467–471. (in Chinese)

Wang QW, Lin M, Zeng M, Tian L (2008) Computational analysis of heat transfer and pressure drop performance for internally finned tube with three different longitudinal fins. Heat Mass Transf 45:147–156

Wang QW, Lin M, Zeng M (2009) Effect of lateral fin profiles on turbulent flow and heat transfer performance of internally finned tubes. Appl Therm Eng 29(14/15):3006–3013

Webb RL (1987) Enhancement of single-phase heat transfer (Chapter 17). In: Kakac S, Shah RK, Aung W (eds) Handbook of single-phase heat transfer. Wiley, New York, pp 17.1–17.62

Weierman C, Taborek J, Marner WJ (1974) Comparison of the performance of in-line and staggered banks of tubes with segmented fins. AIChE Symp Ser 74(174):39–46

Weyrauch E (1969) The influence of geometry of tube banks on heat transfer and pressure drop, when the fluid is flowing normal to the finned bank. Kaltetechnik-Klimatisieurung 21:62–65

Wu JM, Tao WQ (2007) Numerical computation of laminar natural convection heat transfer around a horizontal compound tube with external longitudinal fins. Heat Transf Eng 28(2):93–102

Yu B, Tao WQ (2004) Pressure drop and heat transfer of turbulent flow in annular tubes with internal wave-like longitudinal fins. Heat Mass Transf 40:643–651

Yu B, Nie JH, Wang QW, Tao WQ (1999) Experimental study on the pressure drop and heat transfer characteristics of tubes with internal wave-like longitudinal fins. Heat Mass Transf 35:65–73

Yudin VF, Tokhtarowa LS (1964) Heat emission and resistance of checkerboard and corndor fin clusters. Energomashinostroenie 1:11–13

Chapter 4
Enhanced Plate Fin Geometries with Round Tubes and Enhanced Circular Fin Geometries

Circular tubes mostly have the wavy or herringbone fin and the offset fin or parallel louvre as the major enhanced surface geometries. Wavy fins typically have 50–70% higher heat transfer coefficient than that of a plain fin. In this case, the combination of tubes and a special surface geometry establishes very complex flow geometry.

Two basic wave fin geometries—smooth wave and herringbone wave—are in use (Fig. 4.1). Number of investigations made with herringbone wave geometry is much more than that with smooth wave geometry.

Tao et al. (2007a, b) numerically investigated the performance of wavy fin used for heat transfer augmentation. They used wavy fin (fin A) and plain fin (fin C) and observed that for both the fins, the local Nusselt number decreased along the length of the pipe. The local Nusselt number upstream was almost ten times that of downstream. Thus, they suggested a new fin pattern in which wave is located only in the upstream and referred to it as fin B. The schematic of fin A, fin B and fin C has been shown in Fig. 4.2.

The Nusselt number variation for fin A, fin B and fin C has been presented in Fig. 4.3. The Nusselt number for fin B was about 45% greater than that of fin C and 4% lower than that of fin A. Figure 4.4 shows the friction factor variation with Reynolds number for all the three fins used in the analysis. The plain fin (fin C) as expected has the minimum friction factor. The pressure drop in the case of fin B was found to be less than that of fin A. The friction factor for fin B was 26% higher and 18% lower than that of fin C and fin A, respectively. The overall performance evaluation index, Nu/f, for these three fins has been plotted and shown in Fig. 4.5. The overall performance of fin B has been observed to be the best among the three fins used for the analysis. The Nu/f for fin B has been reported to be 12.4–18.5% and 14.9–20% greater than that of fin C and fin A, respectively.

Nishimura et al. (1987), Xin and Tao (1988), Patel et al. (1991a, b), Rutledge and Sleicher (1994), Comini et al. (2002), Yoshii et al. (1973), Beecher and Fagan (1987), Mirth and Ramadhyani (1994), Xin et al. (1994), Wang et al. (1997, 2002a, b, c), Somchai and Yutasak (2005), Jang and Chen (1997), Tao et al.

© The Author(s), under exclusive license to Springer Nature Switzerland AG 2020
S. K. Saha et al., *Heat Transfer Enhancement in Externally Finned Tubes and Internally Finned Tubes and Annuli*, SpringerBriefs in Applied Sciences and Technology, https://doi.org/10.1007/978-3-030-20748-9_4

Fig. 4.1 Two basic
geometries of the wavy fin:
(a) herringbone wave, (b)
smooth wave (Webb and
Kim 2005)

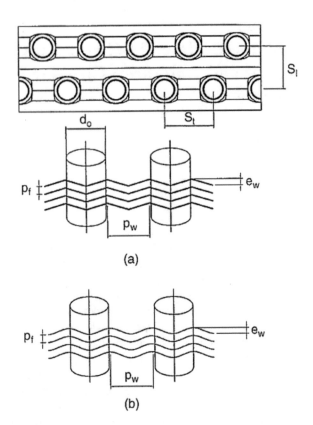

(2007b), Kuan et al. (1984), Zabronsky (1955), Chen and Liou (1998), Saboya and
Sparrow (1974), Jones and Russell (1980), Rosman et al. (1984) and Ay et al. (2002)
have studied the performance of fins.

Goldstein and Sparrow (1977) have observed that, for herringbone wave geometry, enhancement is due to Goetler vortices formed on concave wave surfaces.
Beecher and Fagan (1987) worked with 20, three-row fin-and-tube geometries
having wavy fin geometry. Webb (1990) used their data and developed multiple
regression correlation.

$$Nu_a = 0.5 Gz^{0.86} \left(\frac{S_t}{d_0}\right)^{0.11} \left(\frac{s}{d_0}\right)^{-0.09} \left(\frac{e_w}{S_l}\right)^{0.12} \left(\frac{p_w}{S_l}\right)^{-0.34} \qquad Gz \le 25 \qquad (4.1)$$

$$Nu_a = 0.83 Gz^{0.76} \left(\frac{S_t}{d_0}\right)^{0.13} \left(\frac{s}{d_0}\right)^{-0.16} \left(\frac{e_w}{S_l}\right)^{0.25} \left(\frac{p_w}{S_l}\right)^{-0.43} \qquad Gz > 25 \qquad (4.2)$$

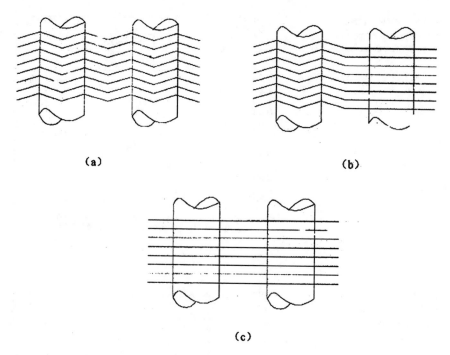

(a)   (b)

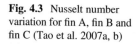

(c)

Fig. 4.2 Schematic of (**a**) fin A, (**b**) fin B and (**c**) fin C (Tao et al. 2007a, b)

Fig. 4.3 Nusselt number
variation for fin A, fin B and
fin C (Tao et al. 2007a, b)

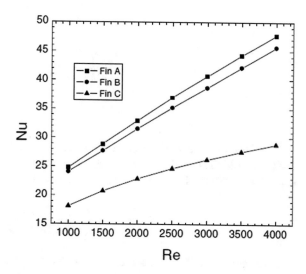

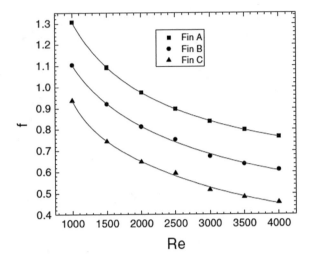

Fig. 4.4 Friction factor variation with Reynolds number (Tao et al. 2007a, b)

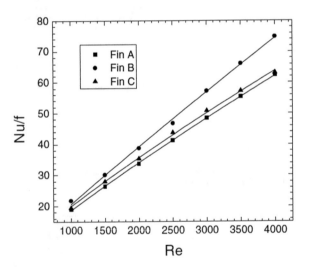

Fig. 4.5 Overall performance evaluation index, *Nu/f* vs. *Re* (Tao et al. 2007a, b)

The Nusselt numbers are sometimes based on arithmetic mean temperature difference (AMTD) and sometimes, as usual, on log mean temperature difference (LMTD). Equation (4.3) gives a relation between Nu_l and Nu_a.

$$Nu_l = \frac{Gz}{4} \ln \left(\frac{1 + 2Nu_a/Gz}{1 - 2Nu_a/Gz} \right) \tag{4.3}$$

Torii and Yang (2007) theoretically studied thermo-hydraulic characteristics of flow over slot-perforated flat fins. The impact of fin pitch on heat transfer and pressure drop characteristics has been elaborated. Similar works have been carried out by Gan et al. (1990), Lee (1995), Biber (1996), de Lieto Vollaro et al. (1999), Anand et al. (1992), Ledezma and Bejan (1996), Leon et al. (2002), Culham and Muzychka (2000) and Furukawa and Yang (2002). Liang and yang (1975a, b), Liang et al. (1977), Lee and Yang (1978), Fujii et al. (1988) and Hwang et al. (1996) studied the effect of perforations on extended surfaces for internal turbine blade cooling.

Bilir et al. (2010) studied the heat transfer and pressure characteristics of fin-tube heat exchanger with three different types of vortex generator configurations. They numerically investigated the effects of location of vortex generators on the heat transfer and pressure drop characteristics. Each type vortex generators were placed at four different locations on the fin to find optimal location so that maximize the heat transfer and minimize the pressure drop. They made two different types of numerical model to optimize the heat transfer characteristics after finding the best location of vortex generators on the fin. They analysed the cumulative effect of three different vortex generators together on the heat transfer rate. After computational analysing and then compared with existing experimental and numerical results in literatures, they found that the use of three different vortex generators together increases heat transfer rate with a moderate increase in pressure drop.

Table 4.1 shows the location of vortex generators and model names. They compared the average heat transfer coefficient results obtained from the numerical analysis with the experimental and computational results of Wu and Tao (2008) as shown in the Fig. 4.6. Table 4.2 shows the numerical results of heat transfer and pressure drop in the heat exchanger. It was found that fin heat transfer rate was ten times the segment heat transfer rate. They concluded that location 4 was best for all the vortex generators in terms of maximum heat transfer rate as well as minimum overall pressure drop. Table 4.3 shows the heat transfer and pressure drop values of the fins with three types of winglet vortex generators. Abu Madi et al. (1998), Chen et al. (2000), Chen and Shu (2004), Elyyan et al. (2008), Kundu and Das (2007), Kwak et al. (2003), Leu et al. (2004), Lozza and Merlo (2001), Méndez et al. (2010), Wang et al. (2002a, b) and Wu and Tao (2008) investigated the effect of fin spacing with vortex generator on heat transfer rate and pressure drop.

Wang et al. (1997, 1998, 1999a, c) and Abu Madi et al. (1998) worked with herringbone wave fin geometry, mostly on staggered layout. The effect of fin pitch and the effect of the rows were studied. General j and f correlations for the herringbone wave configuration were developed by Kim et al. (1997). A procedure in the line of Gray and Webb (1986) was taken for the development of the correlation. Wang et al. (1999d) also developed correlations for the herringbone wave geometry. Kim et al. (1997) correlations are given below:

$$j_3 = 0.394 Re_D{}^{-0.357} \left(\frac{S_t}{S_l}\right)^{-0.272} \left(\frac{s}{d_0}\right)^{-0.205} \left(\frac{e_w}{s}\right)^{-0.133} \left(\frac{p_w}{2e_w}\right)^{-0.558} \tag{4.4}$$

Table 4.1 Model names and location of vortex generators (Bilir et al. 2010)

Location	Distance from the bottom of the fin, d (mm)	Fins with balcony Model name				Fins with imprint Model name				Fins with winglet Model name				Fins with winglet and balcony Model name	Fins with balcony, imprint and winglet Model name
		B1	B2	B3	B4	I1	I2	I3	I4	W1	W2	W3	W4	WBI	WBI
1	6.25	B				I				W				I	B
2	10.875		B				I				W				
3	15.5			B				I				W		B	I
4	24.75				B				I				W	W	W

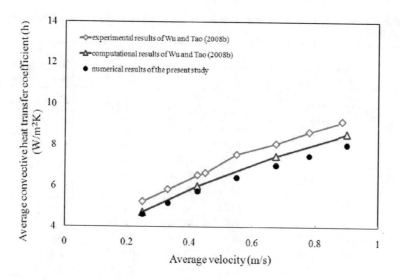

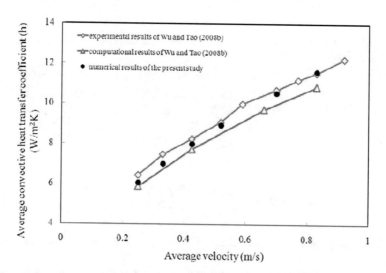

Fig. 4.6 Comparison of the numerical results of the present study with the results of Wu and Tao (2008): (**a**) for a plate fin; (**b**) for the fin with winglet with a 45° angle of attack (Bilir et al. 2010)

$$\frac{j_N}{j_3} = 0.978 - 0.01N \quad Re_d > 1000 \tag{4.5}$$

$$\frac{j_N}{j_3} = 1.35 - 0.162N \quad Re_d < 1000 \tag{4.6}$$

Table 4.2 Numerical results of heat transfer and pressure drop across the heat exchanger (Bilir et al. 2010)

	Model name	$\dot{Q}$ (per segment) (W)	$\dot{Q}$ (per fin) (W)	Normalized $\dot{Q}$ (%)	Total pressure drop (Pa)	Normalized total pressure drop (%)
	P	24.3413	243.413	100	4.0833	100
Fins with balcony	B1	24.4065	244.065	100.267	4.5201	110.697
	B2	24.4390	244.390	100.401	4.7194	115.578
	B3	24.4669	244.669	100.516	4.8049	117.672
	B4	24.6013	246.013	101.068	4.5776	112.105
Fins with imprint	I1	24.3907	243.907	100.203	4.1877	102.557
	I2	24.4116	244.116	100.289	4.2497	104.075
	I3	24.4363	244.363	100.390	4.2850	104.939
	I4	24.4635	244.635	100.502	4.2006	102.872
Fins with winglet	W1	23.7887	237.887	97.730	6.0242	147.532
	W2	23.8291	238.291	97.895	9.3382	228.692
	W3	24.2267	242.267	99.529	9.1347	223.708
	W4	24.5409	245.409	100.820	4.9071	120.174

Table 4.3 Heat transfer and pressure drop values of the fins with three types of vortex generators (Bilir et al. 2010)

Model name	$\dot{Q}$ (per segment) (W)	$\dot{Q}$ (per fin) (W)	Normalized $\dot{Q}$ (%)	Total pressure drop (Pa)	Normalized total pressure drop (%)
P	24.3413	243.413	100	4.0833	100
WBI	24.7808	247.808	101.806	5.7613	141.094
WIB	24.7391	247.391	101.634	5.5524	135.978

$$f_f = 4.467 Re_D^{-0.423} \left(\frac{S_t}{S_l}\right)^{-1.08} \left(\frac{s}{d_0}\right)^{-0.034} \left(\frac{p_w}{2e_w}\right)^{-0.672} \qquad (4.7)$$

Kim et al. (1997) used Zukauskas (1972) correlation for the friction factor due to tubes.

Zhang et al. (2019) used Taguchi method to study the influence of geometric parameters of three-dimensional finned tube on the gas-side heat transfer and pressure drop characteristics in the air cross flow. The effect of four factors such as the fin height, fin width, axial fin pitch and circular fin pitch in compact heat exchanger had been investigated. They developed empirical correlations for Nusselt number and friction factor in the experimental range for the evaluation of heat transfer performance. They found that thermo-hydraulic performance of three-dimensional finned tube was 2.7–2.9 times more than smooth tube in the field of heat transfer criterion. Air and water were used as working medium in the shell side and the tube side, respectively. They worked in heat exchanger with air velocity ranging from 2.9 to 8.3 m/s and the Reynolds number ranging from 4000 to 11,500.

Fig. 4.7 An overview of the changes of (**a**) Nusselt number and (**b**) friction factor with increasing Reynolds number (Zhang et al. 2019)

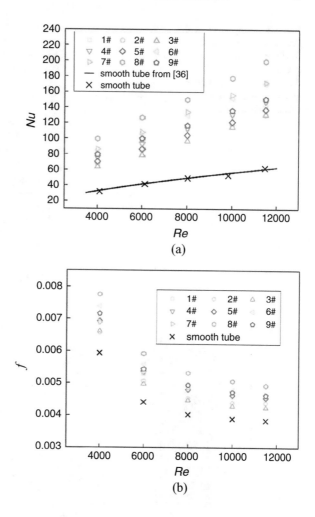

Figure 4.7 shows that coupling effect of different fin parameters on Nusselt number and friction factor. It was cleared from the figure that at the higher value of Reynolds number, heat transfer can be maximized with minimum frictional loss in the finned tube heat exchanger. They experimentally measured that the impact of fin height, axial fin pitch and circular fin pitch on performance evaluation criterion were approximately 47%, 31% and 16%, respectively. Han et al. (2013), Khoshvaght-Aliabadi et al. (2016), Benmachiche et al. (2017), He et al. (2012), Jin et al. (2013), Liao (1990), Anoop et al. (2015), Lemouedda et al. (2012) and Bouzari and Ghazanfarian (2016) had studied about the effect of different types of fins such as circular fins, spiral fins, H-type fins, plate fins, etc. on the hydrothermal performance.

Mirth and Ramadhyani (1994) investigated and developed correlation for smooth wave configuration for the staggered tube layout. Youn et al. (1998) generated data for two-row heat exchangers. Kang and Webb (1998) studied offset strip fins or slit

Fig. 4.8 Comparison of the heat transfer coefficient for the OSF and plain fin geometries for 9.5 mm diameter tubes, 525 fins/m and 0.2 mm fin thickness (Nakayama and Xu 1983)

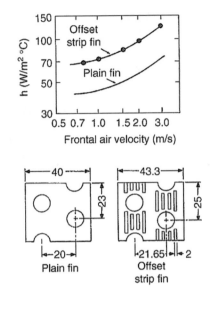

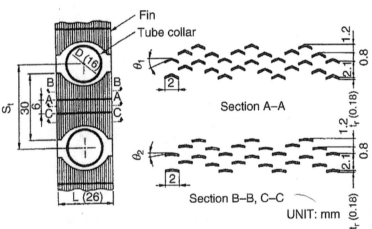

Fig. 4.9 Convex louvre plate fin-and-tube geometry tested (Hatada et al. 1989)

fins applied to fin-tube heat exchangers (Fig. 4.8). Hitachi (1984) used convex louvre fin geometry in commercial plate fin-and-tube heat exchangers. The flow acceleration and fluid mixing in the wake of the tube provide a substantial enhancement. Hatada et al. (1989) generated performance data of the convex louvre fin geometry for a one-row heat exchanger (Figs. 4.9 and 4.10).

The reduced louvre angle near the tube allows more air flow in the vicinity of the tubes. Hatada and Seshu (1984) studied the plate-and-fin geometry. The effect of fin pitch and tube rows on the j and f factors of the convex louvre geometry have been

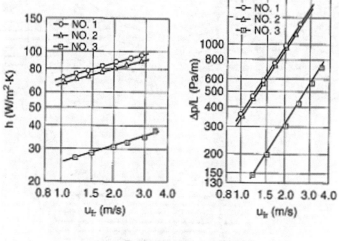

Heat Exchanger Dimensions

	Convex strip fin		Plain plate fin
Feature	Fin no. 1	Fin no. 2	Fin no. 3
Transverse tube pitch S_t (mm)	38	38	36
Fin depth L (mm)	26	26	42
Number of rows	1	1	1
Tube diameter d (mm)	16	16	16
Fin pitch p_t (mm)	2.2	2.2	2.1
Fin thickness t (mm)	0.18	0.18	0.18
Ramp angle θ_1 (Degrees)	17.5	12.5	—
Ramp angle θ_2 (Degrees)	40	12.5	—

Fig. 4.10 Performance data for convex louvre surface geometries (Hatada et al. 1989)

investigated by Wang et al. (1996, 1998). j factors were independent of the fin pitch. The j factors were independent of fin pitch. The row effect on the j factors was relatively weak compared with that of the plain fin geometry. The friction factors were independent of the number of tube rows. The friction factors of the convex louvre fin geometry showed 21–41% and 60–72% increase as compared to the

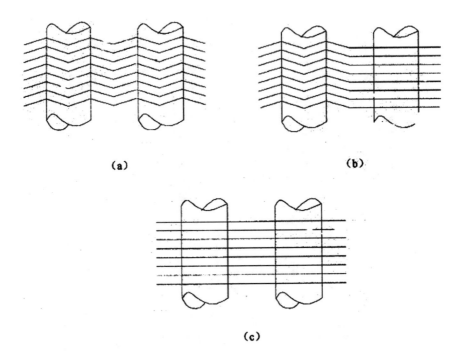

Fig. 4.11 Schematic of (**a**) fin A, (**b**) fin B and (**c**) fin C (Tao et al. 2007a, b)

corresponding wavy fin geometry. Convex louvre geometry performance was the best, followed by the louvre and wavy fin geometries.

Tao et al. (2007a, b) numerically investigated the performance of wavy fin used for heat transfer augmentation. They used wavy fin (fin A) and plain fin (fin C) and observed that for both the fins, the local Nusselt number decreased along the length of the pipe. The local Nusselt number upstream was almost ten times that of downstream. Thus, they suggested a new fin pattern in which wave is located only in the upstream and referred to it as fin B. The schematic of fin A, fin B and fin C has been shown in Fig. 4.11.

The Nusselt number variation for fin A, fin B and fin C has been presented in Fig. 4.12. The Nusselt number for fin B was about 45% greater than that of fin C and 4% lower than that of fin A. Figure 4.13 shows the friction factor variation with Reynolds number for all the three fins used in the analysis. The plain fin (fin C) as expected has the minimum friction factor. The pressure drop in the case of fin B was found to be less than that of fin A. The friction factor for fin B was 26% higher and 18% lower than that of fin C and fin A, respectively. The overall performance evaluation index, Nu/f, for these three fins has been plotted and shown in Fig. 4.14. The overall performance of fin B has been observed to be the best among the three fins used for the analysis. The Nu/f for fin B has been reported to be 12.4–18.5% and 14.9–20% greater than that of fin C and fin A, respectively.

Fig. 4.12 Nusselt number
variation for fin A, fin B and
fin C (Tao et al. 2007a, b)

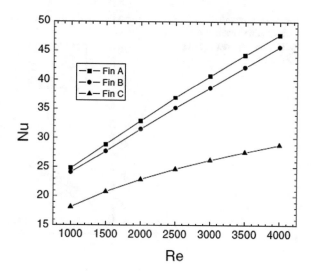

Fig. 4.13 Friction factor
variation with Reynolds
number (Tao et al. 2007a, b)

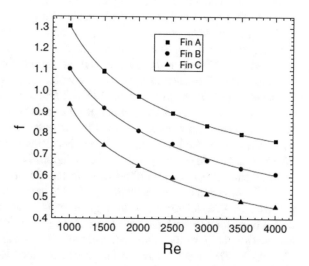

Generalized empirical correlations for *j* and *f* versus *Re* have not been developed for OSF geometry on round tubes. However, Nakayama and Xu developed an empirical correlation. In offset fin geometry, the direction of the strip relative to the air flow direction is very important. Several OSF geometry studies have been done (Wang and Chang 1998; Wang et al. 1999b; Kang and Webb 1998; Yun and Lee 2000; Du and Wang 2000). Youn et al. (2003) investigated the performance of the radial strip geometry. Radial strips perform better than the normal strips since the former have better heat conduction path, and this improves the heat transfer. However, radial strips face the air flow at an oblique angle; this lengthens the effective strip width and slightly reduces the heat transfer.

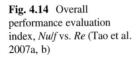

Fig. 4.14 Overall
performance evaluation
index, *Nu/f* vs. *Re* (Tao et al.
2007a, b)

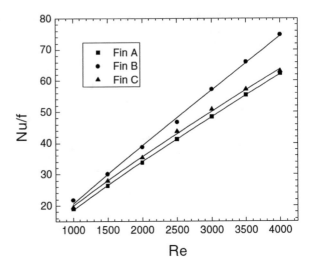

Nakayama and Xu (1983), Kang and Webb (1998), Wang et al. (1999b), Du and Wang (2000) and Youn et al. (2003) dealt with j and f correlations of OSF heat exchangers. However, the applicability of these correlations is very limited. The louvre geometry is applied to fin-tube heat exchangers. The louvre fin must be designed carefully since the louvers can cut the conduction path from the tube. The air-side performance of louvred fin heat exchangers has been investigated by Chang et al. (1995) and Wang et al. (1999a, 1999d). The j factors were independent of fin pitch. The effect of the number of tube rows was negligible for $Re_d > 2000$. However, significant reduction of the j factor with increasing number of tube rows was observed for the lower Reynolds number. Rich (1975) studied the row effect. Wang et al. (1999d) developed j and f correlations based on their data.

Muzychka and Kenway (2009) investigated the performance of offset-strip fin arrays for heat transfer augmentation in liquids having large Prandtl number. They proposed this model for the wake regions of laminar and turbulent regions. The correlations for j factor have been presented to study the effect of Prandtl number suppression on j factor. The schematic of offset-strip fins has been shown in Fig. 4.15. The results have been presented for water, polyalphaolefin and SAE5W30 engine oil considered as working fluids.

Wang et al. (2001) studied slit and louvred fins and observed that performance depends on the louvre or slit pitch, and the fraction of the fin area on which louvres or slit exists. Fujii et al. (1991) studied a plate-fin geometry made of corrugated, perforated plates, and the surface had a one-row having 0.5 mm thick copper fins (Fig. 4.16). The friction performance is not that good compared to the other high-performance fin geometries. The data of the figure are scalable to other tube diameters.

Elyyan and Tafti (2009) investigated the performance of dimpled multilouvred fins for heat transfer augmentation. They used a novel fin configuration with dimples, louvres and perforations. Thus, the combined effect of interrupted surface,

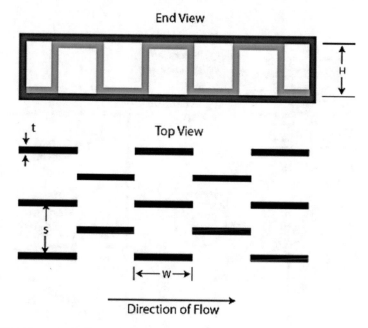

Fig. 4.15 Schematic of offset-strip fins (Muzychka and Kenway 2009)

surface roughness and small-scale discontinuities has been studied. The louvre geometries with dimples have been studied under case 1 and case 2. The fins considered under case 1 have dimples with larger imprint diameter than those in case 2. Case 3 considers perforations on the dimpled louvre fins. The heat transfer enhancement characteristics of fins have been studied by using direct and large eddy simulation.

They observed from the results of case 1 and case 2 that the influence of imprint diameter of the dimples on heat transfer was negligible. The presence of perforation on the dimpled louvre fins redirects the flow from the dimple side to the protrusion side of the fin. Thus, the recirculation in the dimple region is reduced. Further, more flow is observed to be drawn into the dimple cavity resulting in increased vorticity generation. Also, the perforation edges are the regions of high heat transfer coefficients which act as boundary layer regenerators. The flow redirected towards the protruded side and ejecting from there helps in mixing of the flow, and the heat transfer in the wake region of the protrusion is enhanced. They concluded that there was a 12–50% and a maximum 60% increase in heat transfer coefficient and friction factor due to the addition of perforations.

Webb and Trauger (1991), Tafti et al. (1999), Tafti and Zhang (2001), Zhang and Tafti (2001), Lyman et al. (2002), DeJong and Jacobi (2003), Mahmood et al. (2000), Burgess and Ligrani (2004), Ekkad and Nasir (2003), Wang et al. (2003), Ligrani et al. (2001, 2005), Chyu et al. (1997), Moon et al. (2000), Lin et al. (1999), Isaev and Leont'ev (2003), Park et al. (2004), Won and Ligrani (2004), Park and

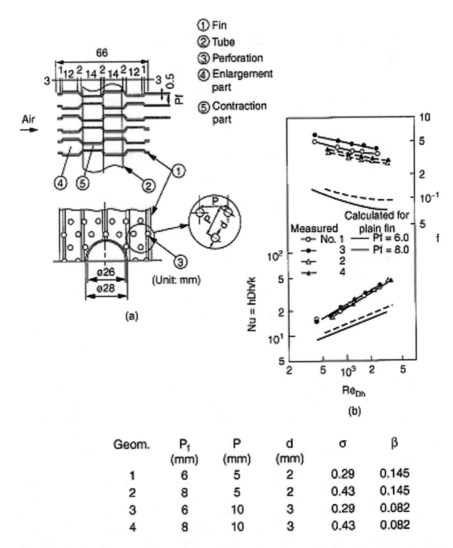

Fig. 4.16 (a) Illustration of one-row fin-tube heat exchanger tested; (b) air-side test results (Fujii et al. 1991)

Ligrani (2005), Patrick and Tafti (2004), Elyyan et al. (2006) and Fujii et al. (1988) studied the heat transfer enhancement using louvred fins.

The mesh fin geometry can be applied to circular fin-tube heat exchangers. Ebisu (1999) observed as much 100% higher heat transfer at the same pumping power than that of conventional louvre fin heat exchangers. Ebisu (1999) extended the work of Torikoshi and Kawabata (1989) for mesh fin heat exchanger with in-line fin configuration, and he investigated the effect of offsetting the fin array. Figure 4.17 shows

Fig. 4.17 Flow visualization results of three-row tube bundles having different tube offsets: (**a**) $y/P = 0$, (**b**) $y/P = 0.1$, (**c**) $y/P = 0.25$, (**d**) $y/P = 0.5$ (Ebisu 1999)

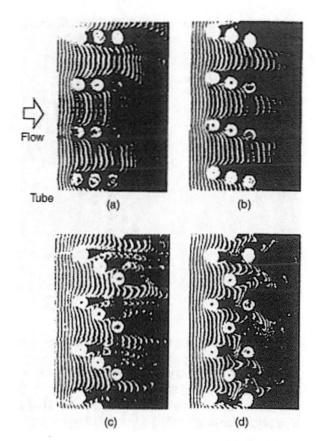

the flow visualization results of three-row tube bundles having different tube offsets. Figure 4.18 compares the performance of copper mesh finned heat exchangers with copper or aluminium louvre fin heat exchangers. The hA/v values of mesh fin having offset fin array and staggered tube layout may be as much as twice that for aluminium louvre fin heat exchanger at the same pumping power.

A low heat transfer coefficient exists in the wake region behind the tubes, particularly at low Reynolds numbers for circular finned tubes. Vortex generators on the fin surface reduce the width of the wake zone and improve heat transfer in the wake region. However, the performance improvement with vortex generators is not great since there is no longitudinal horseshoe vortex which can make significant enhancement on the fins, relative to that provided by vortex generators.

Fiebig et al. (1990) studied vortex generators. The heat transfer enhancement was up to 20% and also the pressure drop decreased up to 10%. This was so because of boundary layer separation on the tube by longitudinal vortices generated by the vortex generators, which give high momentum fluid into the region behind the cylinder. Fiebig et al. (1993) extended their earlier study to three-row heat exchanger

Fig. 4.18 Heat transfer per unit volume (*E*) vs. pumping power per unit volume (*P*) for mesh fin heat exchangers. Louvre fin and plain fin data shown as lines (Ebisu 1999)

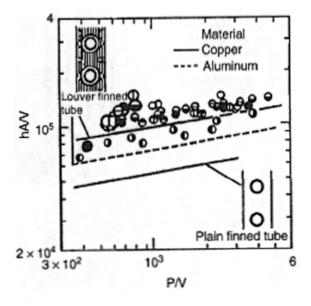

geometry (Fig. 4.19); the vortex generators were in common flow-down configuration.

Torii et al. (2002) investigated the three-row geometry with vortex generators mounted in common flow-up configuration, and they observed pressure loss decrease and minor impairment in heat transfer. They attributed this to the boundary layer separation delay, reduction of form drag and removal of poor heat transfer zone behind the tube.

Kotcioglu and Caliskan (2008) investigated the performance of a cross-flow heat exchanger having wing-type vortex generators. The wing-type vortex generators in particular are convergent-divergent longitudinal vortex generators which are referred to as CDLVGs. An increase in heat transfer rate up to 120% was observed in the heat exchangers due to the presence of vortex generators. They observed a twofold to fourfold increase in pressure drop in case of vortex generators than that in the case of no vortex generators. They evaluated the effectiveness of the cross-flow heat exchanger using the ε-NTU method and observed that the effectiveness was about 60–80% higher in case of heat exchanger with vortex generators than that in the case without CDLVG. The NTU was in the range of 3.32–3.85. The secondary flow was reported in the space between wing cascades due to difference in pressure and velocity which prevails across the space between the converging and diverging pair of winglets.

Garg and Maji (1988) and Maughan and Incropera (1987) used fin, rib and wing configurations as vortex generators. Tahat et al. (2000) presented the spanwise and streamwise fin spacing for in-line and staggered arrangement of fins. El-Sayed et al. (2002) studied the effects of geometrical parameters of the fin, such as fin height, fin thickness, fin spacing, number of fins and fin tip-shroud clearance of fins. Kotcioglu

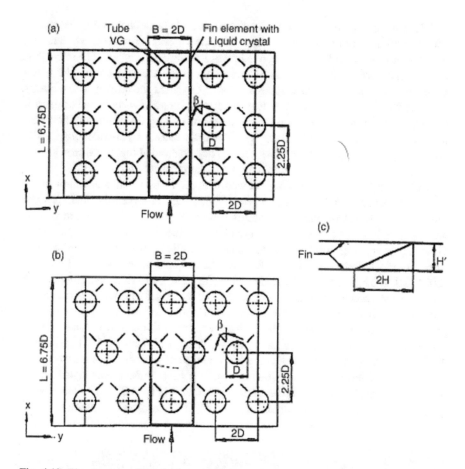

Fig. 4.19 The three-row fin-and-tube geometry tested by Fiebig et al. (1993); (**a**) in-line, (**b**) staggered arrangement, (**c**) shape of the vortex generator, $d = 32$ mm, $H = 7$ mm, 45° angle of attack (Fiebig et al. 1993)

et al. (1998) presented the heat transfer and pressure drop in a rectangular channel using wing-type vortex generators for heat transfer enhancement. Jubran and Al-Salaymeh (1996), Kakaç et al. (1999), Ogulata et al. (2000), Chen and Shu (2004) and Sahin et al. (2005) presented similar work on wing-type vortex generators. Numerical investigation of V-shaped vortex generators has been presented by Sohankar (2007). Tiwari et al. (2003) have also numerically studied the forced convection heat transfer enhancement in a rectangular channel having a built-in oval tube along with delta winglet-type vortex generators. They compared the heat transfer performance using one, two and three winglet vortex generator pairs and concluded that the performance was better for more number of winglet pairs.

Wang et al. (2002a, b, c) studied the fin-and-tube heat exchanger with vortex generators and compared its performance to that of a heat exchanger without vortex generators. They used two types of vortex generators, namely annular winglet type and delta winglet type. The longitudinal vortices have been observed in the case of annular winglet vortex generators. The intensity of counter-rotating vortices was found to increase with the increase in height of the annular winglets. The strength of longitudinal vortices was found to be more intense in case of delta winglet vortex generators. The longitudinal vortices may also be called the stream wise vortices. They explained that the use of vortex generators induces vortices which help in mixing the fluid at the wall with the mainstream flow by disturbing the boundary layer formation at the wall. Also, the form drag caused by slender bodies like winglet-type vortex generators is very less. They concluded that the vortex generators enhanced the heat transfer rate with moderate pressure drop.

Mittal and Balachandar (1995) showed that production and orientation of the vortices depends on the vortex generator types. They observed that the spanwise vortices are oriented parallel to the vortex generator axis. The longitudinal vortices are those which orient along the direction of the flow (Chen et al. 1998). Grossegorgemann et al. (1995) experimentally studied pin-fin array performance. The numerical study on unsteady flow has been taken up by Saha and Acharya (2003, 2004a, b). They used pin-fin arrays for heat transfer enhancement. They observed the enhancement due to three effects: increased surface area due to the presence of fins, boundary layer interruption and enhanced mixing due to the presence of vortex generators which cause vortex shedding and secondary flow. Amon et al. (1992) studied oscillatory flows. Wang and Vanka (1995) and Zhang et al. (1997) have also investigated the performance of periodic pin-fin arrays.

Lozza and Merlo (2001) investigated two-row fin-tube heat exchangers having various enhanced geometries and vortex generators (Fig. 4.20). The addition of winglet vortex generators to louvre fin geometry is not as good as giving the same

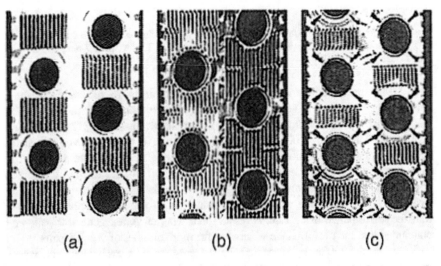

Fig. 4.20 Fin configurations tested by Lozza and Merlo (2001): (**a**) louvre fin A, (**b**) louvre fin B, (**c**) louvre fin with vortex generator (Lozza and Merlo 2001)

area to louvres. Lozza and Merlo (2001) got greater enhancement than that found by Fiebig et al. (1990), who used vortex generators in the tube wake region. Vortex generators do not provide greater enhancement than that can be obtained from conventional slit or louvre fin geometries, when applied to round tubes. Advanced fin geometries reduce the fin efficiency by cutting the fins to form louvres, slits, vortex generators, etc. O'Brien et al. (2003) tested four-row individually finned bundles having annular fins (Fig. 4.21).

Figures 4.22 and 4.23 show enhanced circular fin geometries and segmented or spine fin geometries used in air-conditioning applications, respectively (Webb 1980). All geometries provide enhancement by the periodic development of thin

Fig. 4.21 Individual fins having a pair of winglet vortex generators: (**a**) common flow-down, (**b**) common flow-up configuration (O'Brien et al. 2003)

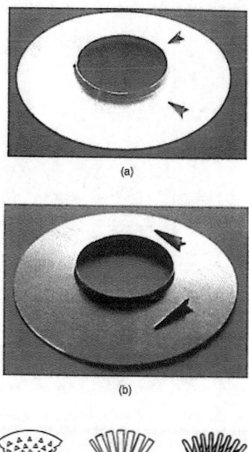

(a)

(b)

(a)   (b)   (c)   (d)   (e)

Fig. 4.22 Enhanced circular fin geometries: (**a**) plain circular fin; (**b**) slotted fin; (**c**) punched and bent triangular projections; (**d**) segmented fin; (**e**) wire loop extended surface (Webb 1987)

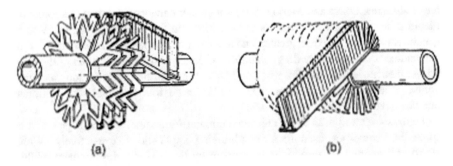

Fig. 4.23 Segmented or spine fin geometries used in air-conditioning applications. (**a**) From La Porte et al. (1979). (**b**) Described by Abbott et al. (1980) and tested by Eckels and Rabas (1985)

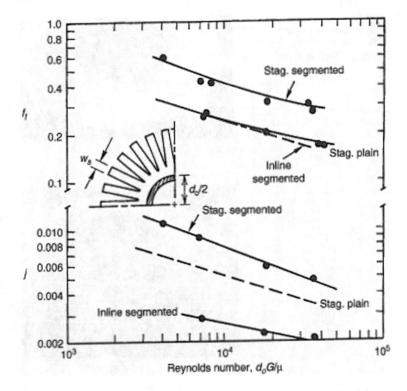

Fig. 4.24 Comparison of segmented fins (staggered and in-line tube layouts) with plain, staggered fin tube geometry as reported: $S_t/d_o = 2.25$, $e/d_o = 0.51$, $s/e = 0.12$, $w/e = 0.17$ (Weierman et al. 1978)

boundary layers on small diameter wires or flat strips, followed by their dissipation in the wave region between elements. The segmented fin is used in a wide range of applications.

Figure 4.24 shows j and f versus Re_d curves for a four-row staggered and a seven-row in-line tube segmented fin geometry (Weierman et al. 1978). It also shows the

j and f curves for a staggered plain fin geometry having the same geometrical parameters as the staggered segmented geometry. Weierman (1976), Rabas et al. (1986) and Breber (1991) studied steel segmented and plain fin geometries for staggered and in-line tube layouts. Steel fin geometries are used for high-temperature applications like boiler economizers and heat recovery boilers to avoid corrosion by combustion products. Breber (1991) also recommended appropriate correlations to predict the heat transfer coefficient and friction factor.

Holtzapple and Carranza (1990) and Holtzapple et al. (1990) studied spine fin tube made of copper tubes and fins. The fins are integral to the tube wall, but these are expensive. Data are provided on several tube pitch layouts. Carranza and Holtzapple (1991) gave an empirical pressure drop correlation. Benforado and Palmer (1964) worked with wire loop fin geometry. They also studied plain circular fin geometry having the same fin pitch and height, and they observed 50% increase in heat transfer coefficient and the same pressure drop as the plain fin.

References

Abbott RW, Norris RH, Spofford WA (1980) Compact heat exchangers in general electric products—sixty years of advances in design and in manufacturing technologies. In: Shah RK, McDonald CF, Howard CP (eds) Compact heat exchangers—history, technology, manufacturing technologies. ASME Symp. HTD, vol 10, pp 37–56

Abu Madi M, Johns RA, Heikal MR (1998) Performance characteristics correlation for round tube and plate finned heat exchangers. Int J Refrig 21(7):507–517

Amon CH, Majumdar D, Herman CV, Mayinger F, Mikic BB, Sekulic DP (1992) Numerical and experimental studies of self-sustained oscillatory flows in communicating channels. Int J Heat Mass Transf 35(11):3115–3129

Anand NK, Kim SH, Fletcher LS (1992) The effect of plate spacing on free convection between parallel plates. J Heat Transf 114:515–527

Anoop B, Balaji C, Velusamy K (2015) A characteristic correlation for heat transfer over serrated finned tubes. Ann Nucl Energy 85(8):1052–1065

Ay H, Jang JY, Yeh JN (2002) Local heat transfer measurements of plate finned-tube heat exchangers by infrared thermography. Int J Heat Mass Transf 45:4069–4087

Beecher DT, Fagan TJ (1987) Effects of fin pattern on the air side heat transfer coefficient in plate finned-tube heat exchangers. ASHRAE Trans 93(2):1961–1984

Benforado DM, Palmer J (1964) Wire loop finned surface—a new application (heat sink for silicon rectifiers). Chem Eng Prog Symp Ser 61(57):315–321

Benmachiche AH, Tahrour F, Aissaoui F, Aksas M, Bougriou C (2017) Comparison of thermal and hydraulic performances of eccentric and concentric annular-fins of heat exchanger tubes. Heat Mass Transf 53(8):2461–2471

Biber CR (1996) Applying computational fluid dynamics to heat sink design and selection. J Electron Cooling 2:22–25

Bilir L, Ozerdem B, Erek A, İlken Z (2010) Heat transfer and pressure drop characteristics of fin-tube heat exchangers with different types of vortex generator configurations. J Enhanc Heat Transf 17(3):243–256

Bouzari S, Ghazanfarian J (2016) Unsteady forced convection over cylinder with radial fins in cross flow. Appl Therm Eng 112:214–225

Breber G (1991) Heat transfer and pressure drop of stud finned tubes. AIChE Symp Ser 87 (283):383–390

Burgess NK, Ligrani PM (2004) Effects of dimple depth on Nusselt numbers and friction factors for internal cooling in a channel. In: Proceedings of ASME Turbo Expo 2004 power for land, sea, and air, GT2004-54232

Carranza RG, Holtzapple MT (1991) A generalized correlation for pressure drop across spined pipe in cross-flow, part I. ASHRAE Trans 97(Pt. 2):122–129

Chang WR, Wang CC, Tsi WS, Shyu RJ (1995) Air-side performance of louver fin heat exchanger. In: Fletcher LS, Aihara T (eds) Proceedings of the ASME/JSME thermal engineering joint conference, vol 4, pp 367–372

Chen HT, Liou JT (1998) Optimum dimensions of the continuous plate fin for various tube arrays. Numer Heat Transf A 34:151–167

Chen TY, Shu HT (2004) Flow structures and heat transfer characteristics in fan flows with and without delta-wing vortex generators. Exp Thermal Fluid Sci 28:273–282

Chen Y, Fiebig M, Mitra NK (1998) Conjugate heat transfer of a finned oval tube, part B: heat transfer behaviours. Numer Heat Transf A 33(4):387–401

Chen Y, Fiebig M, Mitra NK (2000) Heat transfer enhancement of finned oval tubes with staggered punched longitudinal vortex generators. Int J Heat Mass Transf 43:417–435

Chyu MK, Yu Y, Ding H, Downs JP, Soechting FO (1997) Concavity enhanced heat transfer in an internal cooling passage. ASME Paper No. 97-GT-437

Comini G, Nonino C, Savino S (2002) Convective heat and mass transfer in wavy finned-tube exchangers. Int J Numer Meth Heat Fluid Flow 12(6):735–755

Culham JR, Muzychka YS (2000) Optimization of plate fin heat sinks using entropy generation minimization. In: 2000 international society conference on thermal phenomena, Las Vegas, pp 8–15

DeJong NC, Jacobi AM (2003) Localized flow and heat transfer interactions in louvered-fin arrays. Int J Heat Mass Transf 46:443–455

de Lieto Vollaro A, Grignaffini S, Gugliermetti F (1999) Optimum design of vertical rectangular fin arrays. Int J Therm Sci 38:525–529

Du Y-J, Wang C-C (2000) An experimental study of the airside performance of the superslit fin-and-tube heat exchangers. Int J Heat Mass Transf 43:4475–4482

Ebisu T (1999) Evaporation and condensation heat transfer enhancement for alternative refrigerants used in air-conditioning machines. In: Kakac S, Bergles AE, Mayinger F, Yuncu H (eds) Heat transfer enhancement of heat exchangers. Kluwer Academic Publishers, Dordrecht, pp 579–600

Eckels PW, Rabas TJ (1985) Heat transfer and pressure drop performance of finned tube bundles. J Heat Transf 107:205–213

Ekkad SV, Nasir H (2003) Dimple enhanced heat transfer in high aspect ratio channels. J Enhanc Heat Transf 10:395–406

El-Sayed SA, Mohamed MS, Abdel-Latif AM, Abouda AE (2002) Investigation of turbulent heat transfer and fluid flow in longitudinal rectangular-fin arrays of different geometries and shrouded fin array. Exp Thermal Fluid Sci 26:879–900

Elyyan M, Rozati A, Tafti D (2006) Study of flow structures and heat transfer in parallel fins with dimples and protrusions using large eddy simulation. In: Proc. ASME joint US-European fluids engineering summer meeting, Paper No. FEDSM2006-98113

Elyyan MA, Rozati A, Tafti DK (2008) Investigation of dimpled fins for heat transfer enhancement in compact heat exchangers. Int J Heat Mass Transf 51:2950–2966

Elyyan MA, Tafti DK (2009) Flow and heat transfer characteristics of dimpled multilouvered fins. J Enhanc Heat Trans 16(1):43–60

Fiebig M, Mitra NK, Dong Y (1990) Simultaneous heat transfer enhancement and flow loss reduction on fin-tubes. In: Heat transfer 1990. Proceedings of the 9th international heat transfer conference, Jerusalem, vol 4, pp 51–56

Fiebig M, Valencia A, Mitra NK (1993) Wing-type vortex generators for fin-and-tube heat exchangers. Exp Therm Fluid Sci 7(4):287–295

Fujii M, Sehimo Y, Yamanak G (1988) Heat transfer and pressure drop of perforated surface heat exchanger with passage enlargement and contraction. Int J Heat Mass Transf 31(1):135–142

Fujii M, Seshimo Y, Yoshida T (1991) Heat transfer and pressure drop of tube-fin heat exchanger with trapezoidal perforated fins. In: Lloyd JR, Kurosake Y (eds) Proceedings of the 1991 ASME-JSME joint thermal engineering conference, vol 4. ASME, New York, pp 355–360

Furukawa Y, Yang WJ (2002) Reliability of heat sink optimization using entropy generation minimization. In: 8th AIAA/ASME joint thermo-physics and heat transfer conference, St. Louis, Missouri, AIAA-2002-3216 1–6

Gan YP, Lei DH, Wang S (1990) Enhancement of forced convection air cooling of block-like electronic components in in-line arrays. In: Bergles AE (ed) Heat transfer in electronic and microelectronic equipment. Hemisphere Publ. Corp, New York, pp 223–233

Garg VK, Maji PK (1988) Laminar flow and heat transfer in a periodically converging-diverging channel. Int J Numer Meth Fluids 8:579–597

Goldstein L Jr, Sparrow EM (1977) Heat/mass transfer characteristics for flow in a corrugated wall channel. J Heat Transf 99:187–195

Gray DL, Webb RL (1986) Heat transfer and friction correlations for plate fin-and-tube heat exchangers having plain fins. In: Heat transfer 1986. Proceedings of the eighth international heat transfer conference, pp 2745–2750

Grossegorgemann A, Weber D, Fiebig M (1995) Experimental and numerical investigation of self-sustained oscillations in channels with periodic structures. Exp Thermal Fluid Sci 11 (3):226–233

Han H, He YL, Li YS, Wang Y, Wu M (2013) A numerical study on compact enhanced fin-and-tube heat exchangers with oval and circular tube configurations. Int J Heat Mass Transf 65(5):686–695

Hatada T, Senshu T (1984) Experimental study on heat transfer characteristics of convex louver fins for air conditioning heat exchangers. ASME paper ASME 84-HT-74, New York

Hatada D, Ueda U, Oouchi T, Shimizu T (1989) Improved heat transfer performance of air coolers by strip fins controlling air flow distribution. ASHRAE Trans 95(Pt. 1):166–170

He FJ, Cao WW, Yan P (2012) Experimental investigation of heat transfer and flowing resistance for air flow cross over spiral finned tube heat exchanger. In: 2012 international conference on future electrical power and energy system, vol 17 (Part A), pp 741–749

Hitachi Cable Ltd (1984) Hitachi high-performance heat transfer tubes. Cat. No. EA-500. Hitachi Cable, Ltd., Tokyo, Japan

Holtzapple MT, Carranza RG (1990) Heat transfer and pressure drop of spined pipe in cross flow part III: air-side performance comparison to other heat exchangers. ASHRAE Trans 96(Part 2):136–141

Holtzapple MT, Allen AL, Lin K (1990) Heat transfer and pressure drop of spined pipe in cross flow. Part II: heat transfer studies. ASHRAE Trans 96(Part 2):130–135

Hwang GJ, Wu CC, Lin LC, Yang WJ (1996) Investigation of flow drag and forced convective heat transfer in perforated coolant channels. In: Transport phenomena in combustion, vol 2, Taylor & Francis. pp 1747–1758

Isaev SA, Leont'ev AI (2003) Numerical simulation of vortex enhancement of heat transfer under conditions of turbulent flow past a spherical dimple on the wall of a narrow channel. High Temp 41(5):655–679

Jang JY, Chen LK (1997) Numerical analysis of heat transfer and fluid flow in a three- dimensional wavy-fin-and-tube heat exchanger. Int J Heat Mass Transf 40(16):3981–3990

Jin Y, Tang GH, He YL, Tao WQ (2013) Parametric study and field synergy principle analysis of H-type finned tube bank with 10 rows. Int J Heat Mass Transf 60(1):241–251

Jones TV, Russell CMB (1980) Efficiency of rectangular fins. In: Proc. 19th ASME/AIChE national heat transfer conference, Orlando, Florida, pp 27–30

Jubran BA, Al-Salaymeh AS (1996) Heat-transfer enhancement in electronic modules using ribs and "filmcooling-like" techniques. Int J Heat Fluid Flow 17:148–154

Kakaç S, Bergles AE, Mayinger F, Yuncu H (1999) Heat transfer enhancement of heat exchangers, vol 1. Kluwer Academic Publishers, Dordrecht, the Netherlands, pp 75–105

Kang HC, Webb RL (1998) Evaluation of the wavy fin geometry used in air cooled finned tube heat exchangers. In: Heat transfer 1998. Proceedings of the 11th international heat transfer conference, Kyongju, Korea, vol 6, pp 95–100

Khoshvaght-Aliabadi M, Khoshvaght M, Rahnama P (2016) Thermal-hydraulic characteristics of plate-fin heat exchangers with corrugated/vortex-generator plate-fin (CVGPF). Appl Therm Eng 98:690–701

Kim NH, Yun JH, Webb RL (1997) Heat transfer and friction correlations for wavy plate fin-and-tube heat exchangers. Journal of heat transfer, 119(3), 560–567

Kotcioglu I, Caliskan S (2008) Experimental investigation of a cross-flow heat exchanger with wing-type vortex generators. J Enhanc Heat Transf 15(2):113–127

Kotcioglu I, Ayhan T, Olgun H, Ayhan B (1998) Heat transfer and flow structure in a rectangular channel with wing-type vortex generator. Tr J Eng Environ Sci 22:185–195

Kuan DY, Aris R, Davis HT (1984) Estimation of fin efficiencies of regular tubes arrayed in circumferential fins. Int J Heat Mass Transf 27:148–151

Kundu B, Das PK (2007) Optimum dimensions of plate fins for fin-tube heat exchangers. Int J Heat Fluid Flow 18:530–537

Kwak KM, Torii K, Nishino K (2003) Heat transfer and pressure loss penalty for the number of tube rows of staggered finned-tube bundles with a single transverse row of winglets. Int J Heat Mass Transf 46:175–180

LaPorte GE, Osterkorn CL, Marino SM (1979) Heat transfer fin structure. U. S. Patent 4,143,710

Ledezma G, Bejan A (1996) Heat sinks with sloped plate fins in natural and forced convection. Int J Heat Mass Transf 39:1773–1783

Lee S (1995) Optimum design and selection of heat sinks. IEEE Trans Compon Packaging Manuf Technol 18:812–817

Lee CP, Yang WJ (1978) Augmentation of convective heat transfer from high-porosity perforated surfaces. Heat Mass Transf Toronto 2:589–594

Lemouedda A, Schmid A, Franz E, Breuer M, Delgado A (2012) Numerical investigations for the optimization of serrated finned-tube heat exchangers. Appl Therm Eng 31(8):1393–1401

Leon O, De Mey G, Dick E (2002) Study of the optimal layout of cooling fins in forced convection cooling. Microelectron Reliab 42:1101–1111

Leu JS, Wu YH, Jang JY (2004) Heat transfer and fluid flow analysis in plate-fin and tube heat exchangers with a pair of block shape vortex generators. Int J Heat Mass Transf 47:4327–4338

Liang CY, Yang WJ (1975a) Heat transfer and friction loss performance of perforated heat exchanger surface. ASME Heat Transf Conf 97:9–15

Liang CY, Yang WJ (1975b) Modified single blow technique for performance evaluation on heat transfer surface. ASME Heat Transf Conf 97:16–21

Liang CY, Yang WJ, Hung YY (1977) Perforated-fin type compact heat exchangers for gas turbines. 1977 Tokyo Joint Gas Turbine Congress, pp 77–85

Liao G (1990) Experimental investigation of pressure drop and heat transfer of three dimensional internally finned tubes. J Eng Thermophys 04:422–425. (in Chinese)

Ligrani PM, Harrison JL, Mahmood GI, Hill ML (2001) Flow structure due to dimple depressions on a channel surface. Phys Fluids 13(11):3442–3451

Ligrani P, Burgess N, Won S (2005) Nusselt numbers and flow structure on and above a shallow dimpled surface within a channel including effects of inlet turbulence intensity level. J Turbomach 127:321–330

Lin YL, Shih TIP, Chyu MK (1999) Computations of flow and heat transfer in a channel with rows of hemispherical cavities. In: Proc ASME Turbo Expo 1999:99-GT-263

Lozza G, Merlo U (2001) An experimental investigation of heat transfer and friction losses of interrupted and wavy fins for fin-and-tube heat exchangers. Int J Refrig 24:409–416

Lyman AC, Stephen RA, Thole KA, Zhang LW, Memory SB (2002) Scaling of heat transfer coefficients along louvered fins. Exp Thermal Fluid Sci 26:547–563

Mahmood GI, Hill ML, Nelson DL, Ligrani PM (2000) Local heat transfer and flow structure on and above a dimpled surface in a channel. In: Proc ASME Turbo Expo 2000, Paper No. 2000-GT-230

Maughan JR, Incropera PP (1987) Experiments on mixed convection heat transfer for airflow in a horizontal and inclined channel. Int J Heat Mass Transf 30:1307–1318

Méndez RR, Sen M, Yang KY, McClain R (2010) Effect of fin spacing on convection in a plate fin and tube heat exchanger. Int J Heat Mass Transf 43:39–51

Mirth DR, Ramadhyani S (1994) Correlation for predicting the air-side Nusselt numbers and friction factor in chilled water cooling coils. Exp Heat Transf 7:143–162

Mittal R, Balachandar S (1995) Effect of three-dimensionality on the lift and drag of nominally two-dimensional cylinders. Phys Fluids 7:1841–1865

Moon HK, O'Connell TO, Glezer B (2000) Channel height effect on heat transfer and friction in a dimpled passage. J Eng Gas Turbine Power 122:307–313

Muzychka YS, Kenway G (2009) A model for thermal-hydraulic characteristics of offset strip fin arrays for large Prandtl number liquids. J Enhanc Heat Transf 16(1):73–92

Nakayama W, Xu LP (1983) Enhanced fins for air-cooled heat exchangers—heat transfer and friction factor correlations. In: Proceedings of the 1983 ASME-JSME thermal engineering conference, vol 1, pp 495–502

Nishimura T, Yoshino T, Kawamura Y (1987) Numerical flow analysis of pulsatile flow in a channel with symmetric wavy walls at moderate Reynolds numbers. J Chem Eng Japan 20 (5):479–485

O'Brien JE, Sohal MS, Wallstedt PC (2003) Heat transfer testing of enhanced finned tube bundles using the single blow technique. In: Proceedings of the 2003 ASME summer heat transfer conference, Nevada, USA, HT2003-47426

Ogulata RT, Doba F, Yílmaz T (2000) Irreversibility analysis of cross-flow heat exchangers. Energy Convers Manag 41:1585–1599

Park J, Ligrani PM (2005) Numerical predictions of heat transfer and fluid flow characteristics for seven different dimpled surfaces in a channel. Numer Heat Transf A 47:209–232

Park J, Desam PR, Ligrani PM (2004) Numerical predictions of flow structure above a dimpled surface in a channel. Numer Heat Transf A 45:1–20

Patel VC, Chon JT, Yoon JY (1991a) Laminar flow over wavy walls. ASME Trans 113:574–578

Patel VC, Chon JT, Yoon JY (1991b) Turbulent flow over wavy walls. J Fluids Eng 113:578–583

Patrick WV, Tafti DK (2004) Computations of flow structures and heat transfer in a dimpled channel at low to moderate Reynolds number. In: Proc. 2004 ASME heat trans/fluids engineering summer conference, Paper No. HT-FED2004-56171

Rabas TJ, Myers GA, Eckels PW (1986) Comparison of the thermal performance of serrated high-finned tubes used in heat-recovery systems. In: Chiou JP, Sengupta S (eds) Heat transfer in waste heat recovery and heat rejection systems. ASME Symp. HTD, vol 59, pp 33–40

Rich DG (1975) Effect of the number of tube rows on heat transfer performance of smooth plate fin-and-tube heat exchangers. ASHRAE Trans 81(Part 1):307–319

Rosman EC, Carajilescov P, Saboya FEM (1984) Performance of one- and two-row tube and plate fin heat exchangers. ASME J Heat Transf 106:627–632

Rutledge J, Sleicher CA (1994) Direct simulation of turbulent flow and heat transfer in a channel. Part II: a Green's function technique for wavy walls. Commun Numer Meth Eng 10:489–496

Saboya FEM, Sparrow EM (1974) Local and average heat transfer coefficients for one- row plate fin- and tube-heat exchanger configurations. ASME J Heat Transf 96:265–272

Saha AK, Acharya S (2003) Parametric study of unsteady flow and heat transfer in a pin-fin heat exchanger. Int J Heat Mass Transf 46:3815–3830

Saha AK, Acharya S (2004a) Unsteady flow and heat transfer in parallel-plate heat exchangers with in-line and staggered array of posts. Numer Heat Transf A 45:101–133

Saha AK, Acharya S (2004b) Unsteady simulation of turbulent flow and heat transfer in a channel with periodic array of cubic pin-fins. Numer Heat Transf A 46:731–763

Sahin B, Yakut K, Kotcioglu I, Çelik C (2005) Optimum design parameters of a heat exchanger. Appl Energy 82:90–106

Sohankar A (2007) Heat transfer augmentation in a rectangular channel with a V-shaped vortex generator. Int J Heat Fluid Flow 28:306–317

Somchai W, Yutasak C (2005) Effect of fin pitch and number of tube rows on the air side performance of herringbone wavy fin and tube heat exchangers. Energy Convers Manag 46:2216–2223

Tafti DK, Zhang X (2001) Geometry effects on flow transition in multilouvered fins—onset, propagation, and characteristic frequencies. Int J Heat Mass Transf 44:4195–4210

Tafti DK, Zhang LW, Wang G (1999) Time-dependent calculation procedure for fully developed and developing flow and heat transfer in louvered fin geometries. Numer Heat Transf A 35:225–249

Tahat M, Kodah ZH, Jarrah BA, Probert SD (2000) Heat transfers from pin-fin arrays experiencing forced convection. Appl Energy 67(4):419–442

Tao YB, He YL, Wu ZG, Tao WQ (2007a) Numerical design of an efficient wavy fin surface based on the local heat transfer coefficient study. J Enhanc Heat Transf 14(4):315–332

Tao YB, He YL, Huang J, Wu ZG, Tao WQ (2007b) Three-dimensional numerical study of wavy fin-and-tube heat exchangers and field synergy principle analysis. Int J Heat Mass Transf 50:1163–1175

Tiwari S, Maurya D, Biswas G, Eswaran V (2003) Heat transfer enhancement in cross-flow heat exchangers using oval tubes and multiple delta winglets. Int J Heat Mass Transf 46:2841–2856

Torii S, Yang WJ (2007) Thermal-fluid transport phenomena over slot-perforated flat fins with heat sink in forced convection environment. J Enhanc Heat Transf 14(2):123–134

Torii K, Kwak K, Nishino K (2002) Heat transfer enhancement and pressure drop for fin-tube bundles with winglet vortex generators. In: Heat transfer 2002. Proceedings of the 12th international heat transfer conference, vol 4, pp 165–170

Torikoshi K, Kawabata K (1989) Heat transfer and flow friction characteristics of mesh finned air-cooled heat exchangers. In: Figliola RS, Kaviany M, Ebadian MA (eds) Convection heat transfer and transport processes, HTD, vol 116, pp 71–77

Wang G, Vanka SP (1995) Convective heat transfer in periodic wavy passages. Int J Heat Mass Transf 38(17):3219–3230

Wang C-C, Chang Y-J, Hsieh Y-C, Lin Y-T (1996) Sensible heat and friction characteristics of plate fin-and-tube heat exchangers having plane fins. Int J Refrig 19(4):223–230

Wang CC, Fu WL, Chang CT (1997) Heat transfer and friction characteristics of typical wavy fin-and-tube heat exchangers. Heat Transf Friction Charact 14:174–186

Wang C-C, Chang C-T (1998) Heat and mass transfer for plate fin-and-tube heat exchangers with and without hydrophilic coating. Int J Heat Mass Transf 41:3109–3120

Wang C-C, Chang Y-P, Chi K-Y, Chang Y-J (1998) A study of non-redirection louver fin-and-tube heat exchanges. J Mech Eng Sci, 212, SAE 17-212-C1-1

Wang C-C, Chang J-Y, Chiou N-F (1999a) Effects of waffle height on the air-side performance of wavy fin-and-tube heat exchangers. Heat Transf Eng 20(3):45–56

Wang C-C, Du Y-J, Chang Y-J, Tao W-H (1999b) Airside performance of herringbone fin-and-tube heat exchangers in wet conditions. Can J Chem Eng 77:1225–1230

Wang C-C, Jang J-Y, Chiou N-F (1999c) Heat transfer and friction correlation for wavy fin-and-tube heat exchangers. Int J Heat Mass Transf 42:1919–1924

Wang C-C, Lee C-J, Chang C-T, Lin S-P (1999d) Heat transfer and friction correlation for compact louvered fin-and-tube heat exchangers. Int J Heat Mass Transf 42:1945–1956

Wang C-C, Lee W-S, Sheu W-J (2001) A comparative study of compact enhanced fin-and-tube heat exchangers. Int J Heat Mass Transf 44:3565–3573

Wang CC, Hwang YM, Lin YT (2002a) Empirical correlations for heat transfer and flow friction characteristics of herringbone wavy fin-and-tube heat exchangers. Int J Refrig 25:673–680

Wang CC, Lo J, Lin YT, Wei CS (2002b) Flow visualization of annular and delta winglet vortex generators in fin-and-tube heat exchanger application. Int J Heat Mass Transf 45:3803–3815

Wang CC, Lo J, Lin YT, Liu MS (2002c) Flow visualization of wave-type vortex generators having inline fin-tube arrangement. Int J Heat Mass Transf 45:1933–1944

Wang Z, Yeo KS, Khoo BC (2003) Numerical simulation of laminar channel flow over dimpled surface. In: Proc. AIAA conference, Paper No. AIAA 2003-3964

Webb RL (1980) Air-side heat transfer in finned tube heat exchangers. Heat Transf Eng 1(3):33–49

Webb RL (1987) Enhancement of single-phase heat transfer (Chapter 17). In: Kakac S, Shah RK, Aung W (eds) Handbook of single-phase heat transfer. Wiley, New York, pp 17.1–17.62

Webb RL (1990) Air-side heat transfer correlations for flat and wavy plate fin-and-tube geometries. ASHRAE Trans 96(Part 2):445–449

Webb RL, Kim NY (2005) Principles of enhanced heat transfer. Taylor & Francis, New York

Webb RL, Trauger P (1991) Flow structure in the louvered fin heat exchanger geometry. Exp Thermal Fluid Sci 4:205–217

Weierman C (1976) Correlations ease the selection of finned tubes. Oil Gas J 74:94–100

Weierman C, Taborek J, Marner WJ (1978) Comparison of the performance of inline and staggered banks of tubes with segmented fins. AIChE Symp Ser 74(174):39–46

Won S, Ligrani P (2004) Numerical predictions of flow structure and local Nusselt number ratios along and above dimpled surfaces with different dimple depths in a channel. Numer Heat Transf A 46:549–570

Wu JM, Tao WQ (2008) Numerical study on laminar convection heat transfer in a channel with longitudinal vortex generator. Part B: parametric study of major influence factors. Int J Heat Mass Transf 51:3683–3692

Xin RC, Li HZ, Kang HJ, Li W, Tao WQ (1994) An experimental investigation on heat transfer and pressure drop characteristics of triangular wavy fin-and-tube heat exchanger surfaces. J Xi'an Jiaotong Univ 28(2):77–83

Xin RC, Tao WQ (1988) Numerical prediction of laminar flow and heat transfer in wavy channels of uniform cross-sectional area. Num Heat Transf 14:465–481

Yoshii T, Yamamoto M, Otaki T (1973) Effects of dropwise condensate on wet heat transfer surface for air cooling oils. In: Proc. 13th international congress of refrigeration, pp 285–292

Youn B, Kil Y-H, Park H-Y, Yoo K-C, Kim Y-S (1998) Experimental study of pressure drop and heat transfer characteristics of 10.07 mm wave and wave-slit fin-tube heat exchangers with wave depth of 2 mm. In: Heat transfer 1998. Proceedings of the 11th international heat transfer conference, vol 6, Kyongju, Korea

Youn B, Kim Y-S, Park H-Y, Kim N-H (2003) An experimental investigation on the airside performance of fin-and-tube heat exchangers having radial slit fins. J Enhanc Heat Transf 10:61–80

Yun J-Y, Lee K-S (2000) Influence of design parameters on the heat transfer and flow friction characteristics of the heat exchanger with slit fins. Int J Heat Mass Transf 43:2529–2539

Zabronsky H (1955) Temperature distribution and efficiency of a heat exchanger using square fins on round tubes. ASME J Appl Mech 22:119–122

Zhang X, Tafti DK (2001) Classification and effects of thermal wakes on heat transfer in multilouvered fins. Int J Heat Mass Transf 44:2461–2473

Zhang LW, Tafti DK, Najjar FM, Balachandar S (1997) Computations of flow and heat transfer in parallel-plate fin heat exchangers on the CM-5: effects of flow unsteadiness and three-dimensionality. Int J Heat Mass Transf 40(6):1325–1341

Zhang JN, Cheng M, Ding YD, Fu Q, Chen ZY (2019) Influence of geometric parameters on the gas-side heat transfer and pressure drop characteristics of three-dimensional finned tube. Int J Heat Mass Transf 133:192–202

Zukauskas A (1972) Heat transfer from tubes in crossflow. In: Hartnett JP, Irvine TF (eds) Advances in heat transfer, vol 8. Academic Press, New York, pp 93–160

Chapter 5
Oval and Flat Tube Geometries, Row Effects in Tube Banks, Local Heat Transfer Coefficient on Plain Fins, Performance Comparison, Numerical Simulation and Patents, Coatings

5.1 Oval and Flat Tube Geometries

Oval and flat cross-sectional tube shapes are also applied to individually finned tubes (Fig. 5.1) (Brauer 1964). These tubes are better due to lower form drag on the tubes and the smaller wake region on the fin behind the tube; but the tube-side design pressure must be sufficiently low. Min and Webb (2004) studied numerically the effect of tube aspect ratio of an oval tube on the air-side heat transfer and pressure drop characteristics of an infinite row heat exchanger having herringbone wavy fins. They used five tube geometries: a round tube, three elliptical oval tubes and a flat tube (Fig. 5.2). The geometric details are given in Table 5.1. Webb and Iyengar (2001) (Fig. 5.3) compared the air-side performance of oval tube geometry with that of a two-row finned tube heat exchanger.

The thermo-hydraulic performance of wavy fin heat exchanger with elliptical tube was investigated by Tao et al. (2007) by using three-dimensional simulation method and field synergy principle. They examined the five important factors which affected the wavy fin and elliptical tube heat exchanger. Numerical results of circular tube were compared with the experimental results of circular and elliptical tubes ($e = b/a = 0.6$) with same minimum cross-sectional area. It was observed from numerical results that the relative heat transfer and corresponding friction factor increased maximum up to 30% and 10%, respectively, in elliptical arrangement. They found an optimum trend of the effective parameters which enhanced the heat transfer rate with some penalty in pressure drop. They observed that heat transfer of finned tube bank can be enhanced with increasing Reynolds number and fin thickness and decreasing eccentricity and spanwise tube pitch with some loss of pressure. They found an optimum fin pitch ($F_b/2b = 0.1$) for efficient heat transfer.

Figure 5.4 shows the detailed analysis of variation of Nusselt numbers and friction factors with respect to Reynolds number and inlet velocity and compared the simulation results of circular and elliptical tube with experimental results

© The Author(s), under exclusive license to Springer Nature Switzerland AG 2020
S. K. Saha et al., *Heat Transfer Enhancement in Externally Finned Tubes and Internally Finned Tubes and Annuli*, SpringerBriefs in Applied Sciences and Technology, https://doi.org/10.1007/978-3-030-20748-9_5

Fig. 5.1 Heat transfer and friction characteristics of circular and oval finned tubes in a staggered tube layout (Brauer 1964)

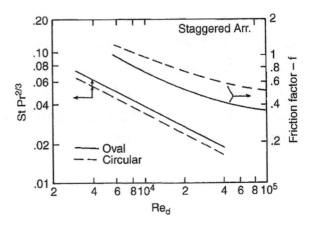

Tube Bank Dimensions (mm)

	Circular	Oval
Tube dia. (d)	29	19.9/35.2
Fin. ht. (e)	9.8	10/9.3
Fin. thk. (t)	0.4	0.4
Face pitch (S_t/d)	1.03	1.05
Row pitch (S_l/d)	1.15	1.04
Fins/m	312	312

Fig. 5.2 Tube geometries considered by Min and Webb for numerical calculation: (**a**) cross-sectional shape, (**b**) computational domain for the ET-2 oval tube case (Min and Webb 2004)

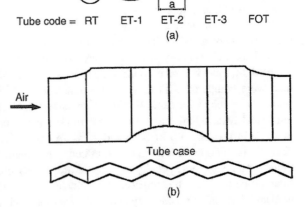

obtained from Xin et al. (1994). The widths of the computational domain for the three cases were same. The air flow cross-sectional area of the elliptic tube A is smaller than that of the elliptic tube B as shown. Figure 5.5 shows the effect of fin pitch on the Nusselt number and friction factor. Wang et al. (1997, 1999), Jang and Chen (1997), Somchai and Yutasak (2005) and Manglik et al. (2005) either

Table 5.1 Tube dimensions for numerical calculations (Min and Webb 2004)

Tube code	a (mm)	b (mm)	a/b	D_h (mm)	$D_h/(D_h)_{RT}$
RT	15.88	15.88	1	14.86	1
ET-1	20.59	10.3	2	12.44	0.82
ET-2	22.37	7.45	3	9.26	0.62
ET-3	23.34	5.44	4.29	6.55	0.45
FOT	20.95	6.98	3	9.54	0.64

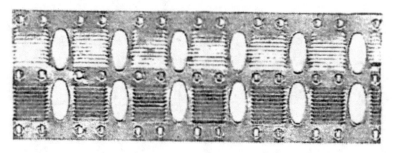

Fig. 5.3 Photo of oval tube fins used in the analysis of Webb and Iyengar (2001)

experimentally or numerically investigated the effect of fin density, fin pitch of herringbone wavy fin, fin pitch, fin height, etc. on the performance of heat exchanger.

O'Conner and Pasternak (1976) worked with flat aluminium extruded tubes with internal membranes. Haberski and Raco (1976), Cox (1973) and Cox and Jallonk (1973) give more information. Webb and Gupte (1990) compared the performance of this heat exchanger construction with the wavy plate fin-and-tube and the spine fin geometries.

Figure 5.6 shows an automotive radiator geometry having louvred plate fins on flat tubes. The automotive radiator operates at low pressure, and therefore, internal membranes are not required in the tubes.

Achaichia and Cowell (1988) gave j and f correlations,

$$\gamma = \frac{1}{\theta} \left(0.936 - \frac{243}{Re_L} - 1.76\frac{P_f}{L_p} + 0.995\theta \right) \tag{5.1}$$

$$j = 1.234\gamma Re_L^{-0.59} \left(\frac{S_t}{L_p} \right)^{-0.09} \left(\frac{p_f}{L_p} \right)^{-0.04} \tag{5.2}$$

$$f = 533 p_f^{-0.22} L_p^{0.25} S_t^{0.26} H^{0.33} \left[Re_L^{0.318\log_{10}Re_L - 2.25} \right]^{1.07} \tag{5.3}$$

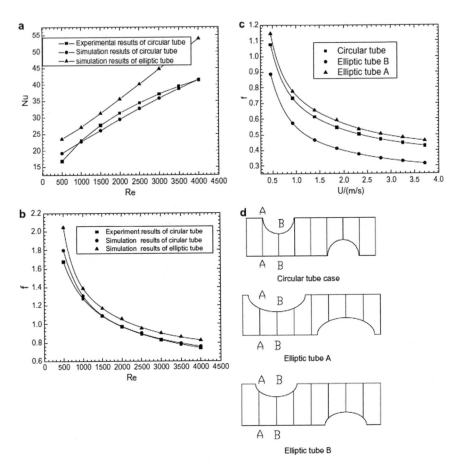

Fig. 5.4 Effects of *Re* on *Nu* and *f*. (**a**) Effect of *Re* on *Nu*; (**b**) effect of *Re* on *f*; (**c**) effect of *Re* on *f* of different tube cases; (**d**) schematic of the flow cross section of the three cases (Tao et al. 2007)

O'Brien et al. (2001) studied vortex geometry on flat and oval fin-tube geometries. Valencia et al. (1996) investigated the effect of vortex generator location
(Fig. 5.7). Fiebig et al. (1994) worked with three-row staggered tube configuration
(Fig. 5.8) and compared the results with those from round tube geometry. The vortex
generators increased the heat transfer by 100% for the flat tube, but only 10% for
the round tube. O'Brien et al. (2001) investigated the effect of vortex generators
on oval tube geometry, and the results are shown in Fig. 5.9. Tests were conducted
in a narrow rectangular duct fitted with an oval tube, and Nusselt numbers for
the round tube geometry without winglets are larger than oval tube geometry without
winglets.

Fig. 5.5 Variations of average Nu and f with fin pitch. (**a**) Variation of average Nu with fin pitch; (**b**) variation of average f with fin pitch (Tao et al. 2007)

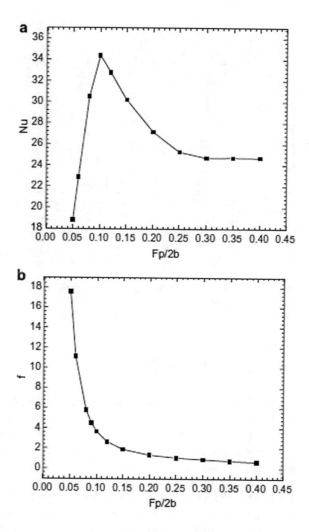

Fig. 5.6 Illustration of the louvred plate fin automotive radiator with inline tubes (Achaichia and Cowell 1988)

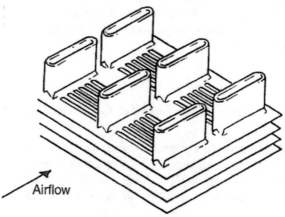

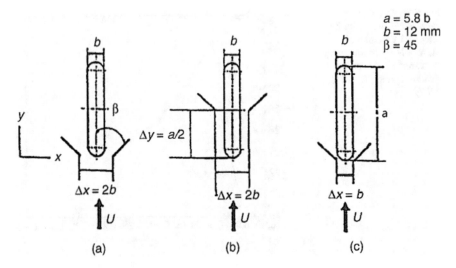

Fig. 5.7 The locations of delta winglet pairs: (**a**) upstream of the tube with the closet winglet spacing equal to twice the tube width, (**b**) at the middle of the tube, (**c**) upstream of the tube with the closet winglet spacing equal to the tube width (Valencia et al. 1996)

5.2 Row Effects in Tube Banks

Most of the available correlations are for deep tube banks, and these do not take care of row effects. In an in-line tube bank, the heat transfer coefficient will decrease with rows due to bypass effects. However, the coefficient increases with number of tube rows in a staggered bank; since turbulent eddies are shed from the tubes and this causes good mixing in the downstream fin region. In-line tube banks generally have a smaller heat transfer coefficient than staggered tube banks (Rabas and Huber 1989). There is a basic difference in the flow phenomena in staggered and in-line finned tube banks. Figure 5.10 compares the performance of in-line and staggered banks of plain, circular finned tubes (Braner 1964). It is argued that bypass effects in the in-line arrangement are responsible for the poor performance; this is evident from Fig. 5.11. Rabas et al. (1986) gave additional data on in-line versus staggered layout for plain tube.

5.3 Local Heat Transfer Coefficient on Plain Fins

The flow accelerates around the tube and forms a wake region behind the tube. This causes local variations of the heat transfer coefficient (Neal and Hitchcock 1966; Jones and Russell 1980; Saboya and Sparrow 1974; Kruckels and Kottke 1970). Saboya and Sparrow (1974, 1976a, b) worked with naphthalene mass transfer. They

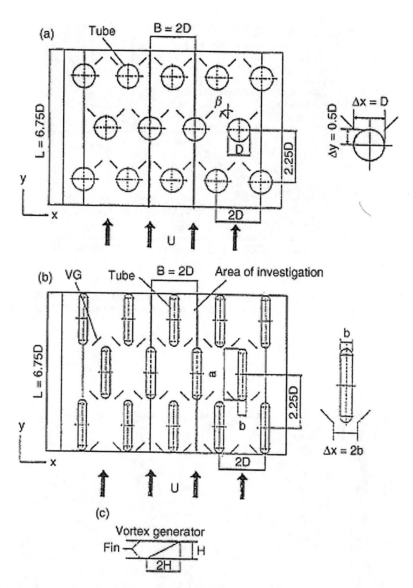

Fig. 5.8 Three-row fin-and-tube geometries tested by Fiebig et al. (1994): (**a**) round tube, (**b**) flat tube, (**c**) shape of the vortex generator, $d = 32$ mm, $H = 7$ mm, $a = 70$ mm, $b = 12$ mm, 45° angle of attack (Webb and Kim 2005)

measured local coefficients for one, two and here row plate fin-and-tube geometries. The analogy drawn by them gave the heat transfer coefficients. Single circular fin-and-tube measurements have been taken by Neal and Hitchcock (1966), Hu and Jacobi (1993), Kearney and Jacobi (1996), Braner (1964) and Kruckels and Kottke

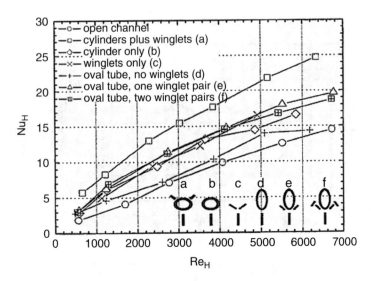

Fig. 5.9 Nu_H vs. Re_H (*H:* channel height) of various geometries tested, $a/H = 8.66$, $d/H = 5.0$, where a is the major diameter of the oval tube and d is the diameter of the circular tube (O'Brien et al. 2001)

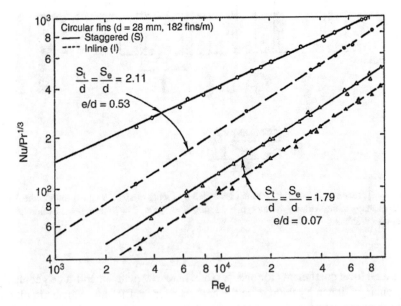

Fig. 5.10 $NuPr$-113 vs. Red for in-line and staggered banks of circular finned tubes with plain fins (Brauer 1964)

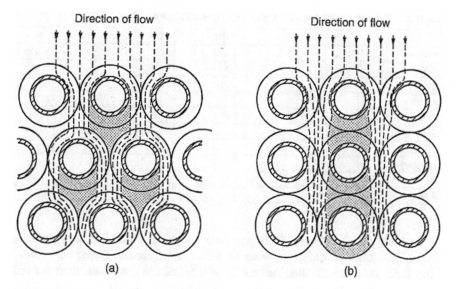

Fig. 5.11 Flow pattern for (**a**) staggered and (**b**) in-line finned tube banks (Brauer 1964)

(1970). They also generated data for banks of staggered arrangement. Following observations have been made:

- Marked higher heat transfer on the upstream area of the fin than on the downstream area. Maximum coefficients occur (70–90) from the forward stagnation point. Higher heat transfer coefficients occur near the fin tip than near the base of the fin.
- Stagnation flow on the front of the tube gives high heat transfer at the fin root. Smaller coefficient results near the fin tip at the front due to flow separation. Radial heat flow occurs only near the front of the fin.

5.4   Performance Comparison, Numerical Simulation and Patents, Coatings

Webb and Gupte (1990) compared the performance of the six enhanced surface geometries as depicted in Table 5.2. Air-side heat transfer and friction characteristics have been analysed, and correlations have been presented by Eckels and Rabas (1985), Mori and Nakayama (1980), Webb (1990), Hatada and Senshu (1984), Wieting (1975) and Davenport (1983). The Kandlikar (1987) correlation predicted the tube-side heat transfer coefficient for vaporization of R-22 in plain tubes. This correlation can also be used for the calculation of heat transfer coefficient for flat tubes using the hydraulic diameter in the Reynolds number definition.

Table 5.2 Heat exchanger geometries compared (Webb and Gupte 1990)

	Spine	Wavy	Slit	CLF	OSF	Louvre
Figure	6.2a	6.2c	6.2b	6.2d	6.35b	6.35c
d_0 (mm)	9.52	9.52	9.52	9.52		
b (nm)					3.46	3.46
a (mm)					7.87	8.38
t (mm)	0.76	0.76	0.76	0.76	0.76	0.76
S_1 (mm)	25.4	23.62	25.4	25.4	21.84	21.84
S_1 (mm)	25.4	20.60	21.60	21.60		
L_p (mm)			1.98		1.59	1.59
n_L			4		5	5
θ (degrees)					20	20

Lindstedt and Karvinen (2012) presented numerical solution to minimize the thermal resistance of isothermal plate fin arrays in the laminar forced convection. They focused on to maximize the heat transfer with some constraints such as fixed number of fins, either volume or width and either pressure drop or fan power. They presented the optimal design of plate fin array by considering the measurement of non-dimensional parameters such as channel length and aspect ratio. They characterized the pressure losses due to expansion and contraction at the inlet and outlet. They solved the governing equations by using analytical methods. They observed volume flow rate as a criterion to enhance the heat transfer rate. Teertstra et al. (2000), Lehtinen (2005), Bejan and Morega (1994), Yilmaz et al. (2000), Muzychka (2005), Canhoto and Heitor Reis (2011), Bejan and Scibba (1992), Bar-Cohen and Rohsenow (1984), and Hetsroni et al. (2011) worked on heat transfer enhancement by using fins in a tube.

Wu and Tao (2007) studied the numerical computation of natural convection heat transfer through horizontal compound tube with external longitudinal fins in laminar region. They analysed the heat transfer of laminar natural convection through compound tube with different fin heights and different numbers of fins by using conjugated computational method with primitive variables. They investigated some parameters which affect the total heat transfer rate: (a) fin height for fixed number of fins, (b) number of fins for fixed fin height and (c) finned tube positioning.

They determined the optimum number of fins, optimum fin height and optimum position of finned tube at which the total heat transfer rate of externally finned tube approached the maximum value. It was observed that the heat transfer rate first profoundly increased with increasing the height of the fins, and then gradually became a constant value when the number of fins was six as shown in the Fig. 5.12. It was found that heat transfer rate first increased with increasing in number of fins then decreased. Finally, they observed that optimum fin number was eight at which the total heat transfer rate reached the maximum and corresponding fin efficiency was approximately 83%.

Saha (2008) investigated the effect of number of periodic modules on heat transfer characteristics of flow in a channel with cubic pin-fin periodic array inside

Fig. 5.12 Influence of relative height of fins on the heat transfer rate of the finned tube with six fins (Wu and Tao 2007)

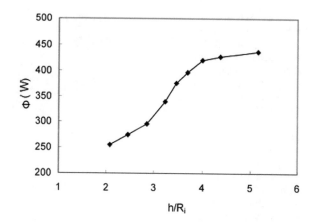

Table 5.3 Comparison of all-aluminium heat exchangers (plain tubes with two refrigerant circuits) (Webb and Kim 2005)

	Spine	Wavy	Slit	CLF	OSF	Louvre
Rows	1	1	1	1	0.59	0.59
Fins/m	728	433	590	433	866	866
u_0 (m/s)	0.96	1.07	1.32	1.19	1.39	1.44
h_e (W/m^2 C)	64.8	81.7	77.7	93.1	130.0	133.1
η	0.93	0.91	0.90	0.89	0.83	0.83
G_{ref} (kg/m^2 s)	336	336	336	336	1127	1107
h_1 (W/m^2 °C)	3531	3554	3690	3667	10,288	9732
w_{fin} (kg)	1.96	1.58	1.85	1.49	0.97	0.93
w_{fuh} (kg)	1.23	1.19	0.96	0.99	0.84	0.84
w_{not} (kg)	3.19	2.77	2.81	2.48	1.81	1.77

it. The unsteady state 3D numerical investigation has been carried out, and the results have been discussed. The in-line arrangement was considered for streamwise and spanwise periodic array of pin fins with periodicity of 2.5 times the dimension of the pin-fin. They have used high-order temporal and spatial filtered-averaged Navier–Stokes equation and Energy equation. The turbulence fluctuations were taken care of by using the Large Eddy Simulation (LES) turbulence model. They presented the results for heat transfer in both instantaneous and time-averaged flow. They observed quite complex flow behaviour. The results of time-averaged Nu agreed well with the experimental results. They concluded that the rms fluctuations of the flow are affected by streamwise and transverse periodic lengths while the time-averaged fields are unaffected.

Table 5.3 compares the various parameters of the six heat exchanger configurations. The spine fin is taken as the reference for comparison. The table gives same heat duty, and each design in Table 6.3 operates at the same air-side pressure drop and airflow rate. The higher pressure drops of the finned-tube heat exchangers have

Fig. 5.13 Patented enhanced fin geometries: (**a**) convex louvre fin by Bemisderfer and Wanner (1991), (**b**) convex louvre fin by Ueda et al. (1994), (**c**) louvre fin by Beamer and Cowell (1998), radial slit fin by Youn and Kim (1998), (**d**) slit fin patented by Yun and Kim (1997) and Jung and Jung (1999), (**e**) vortex generators patented by Esformes (1989), (**f**) according fin patented by Tanaka et al. (1994), (**g**) woven wire fin patented by Ikejima et al. (1998) (Wang 2000)

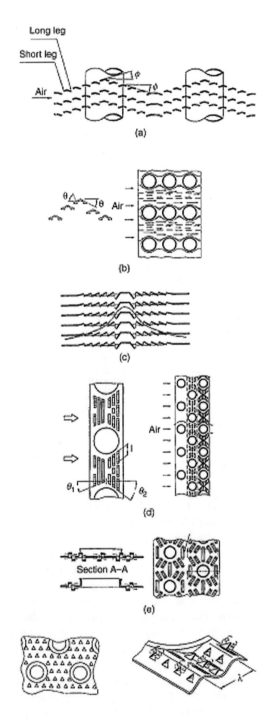

been made by Torikoshi et al. (1994), Torikoshi and Xi (1995), Onishi et al. (1999), Tsai et al. (1999), Jang and Chen (1997), Boewe et al. (1999), Sheui et al. (1999), Leu et al. (2001) and Jang et al. (2001), to name a few investigations. However, all these investigations had been with steady flow. Flow unsteadiness is extremely important in such enhanced geometries. More research efforts need to be directed to the incorporation of flow unsteadiness by complex fin geometries.

Figure 5.13 shows patented enhanced fin geometries. Wang (2000), Bemisderfer and Wanner (1991), Beamer and Cowell (1998), Youn and Kim (1998), Yun and Kim (1997), Jung and Jung (1999), Esformes (1989), Itoh et al. (1986), Tanaka et al. (1994) and Ikejima et al. (1998) have taken international patents on advanced fin geometries.

Condensate forms on evaporator fins when the surface temperature drops below the dew point temperature of the ambient air, and surface wettability is a key parameter for this. Min and Webb (2000), Min et al. (2000), Kim et al. (2002), Mirth and Ramadhyani (1993), McQuiston (1978) and Wang et al. (1997, 2000) have worked with hydrophilic coatings.

References

Achaichia A, Cowell TA (1988) Heat transfer and pressure drop characteristics of flat tube and louvered plate fin surfaces. Exp Therm Fluid Sci 1:147–157

Bar-Cohen A, Rohsenow WM (1984) Thermally optimum spacing of vertical natural convection cooled parallel plates. J Heat Transf 106(1):116–123

Beamer HE, Cowell TA (1998) Heat exchanger cooling fin with varying louver angle. U.S. patent 5,730,214

Bejan A, Morega M (1994) The optimal spacing of a stack of plates cooled by turbulent forced convection. Int J Heat Mass Transf 37(6):1045–1048

Bejan A, Sciubba E (1992) The optimal spacing of parallel plates cooled by forced convection. Int J Heat Mass Transf 35(12):3259–3264

Bemisderfer C, Wanner J (1991) Chevron lanced fin design with unequal leg lengths for a heat exchanger. U.S. patent 5,062,475

Boewe D, Yin J, Park YC, Bullard CW, Hrnjak PS (1999) The role of suction line heat exchanger in transcritical R-744 mobile air-conditioning systems. SAE Int. Congress and Exposition, SAE 1999-01-0583

Brauer H (1964) Compact heat exchangers. Chem Prog Eng (London) 45(8):451–460

Canhoto P, Heitor Reis A (2011) Optimization of forced convection heat sinks with pumping power requirements. Int J Heat Mass Transf 54:1441–1447

Cox B (1973) Heat transfer and pumping power performance in tube banks—finned and bare. ASME Paper 73-HT-27

Cox B, Jallouk PA (1973) Methods for evaluating the performance of compact heat exchanger surfaces. J Heat Transf 95:464–469

Davenport CJ (1983) Correlations for heat transfer and flow friction characteristics of louvered fin. In: Heat transfer—Seattle 1983, AIChE Sym. Ser., No. 225, vol 79, pp 19–27

Eckels PW, Rabas TJ (1985) Heat transfer and pressure drop of typical air cooler finned tubes. J Heat Transf 107:198–204

Esformes JL (1989) Ramp wing enhanced plate fin. U.S. patent 4,817,709

Fiebig M, Valencia A, Mitra NK (1994) Local heat transfer and flow losses in fin-and-tube heat exchangers with vortex generators: a comparison of round and flat tubes. Exp Therm Fluid Sci 8 (1):35–45

Haberski RJ, Raco RJ (1976) Engineering analysis and development of an advanced technology low cost dry cooling tower heat transfer surface. Curtiss-Wright Corporation, Report No.Cod-2774-1

Hatada T, Senshu T (1984) Experimental study on heat transfer characteristics of convex louver fins for air conditioning heat exchangers. ASME paper ASME 84-HT-74

Hetsroni G, Mosyak A, Pogrebnyak E, Yarin LP (2011) Micro-channels: reality and myth. J Fluids Eng 133:121–202

Hu X, Jacobi AM (1993) Local heat transfer behavior and its impact on a single-row, annularly finned tube heat exchanger. J Heat Transf 115:66–74

Ikejima K, Gotoh T, Yumikura T, Takeshita M, Yoshita T (1998) Heat exchanger and method of fabrication the heat exchanger. U.S. patent 5,769,157

Itoh M, Kogure H, Iino K, Ochiai I, Kitayama Y, Miyagi M (1986) Fin-and-tube type heat exchanger. U.S. patent 4593756

Jang JY, Chen LK (1997) Numerical analysis of heat transfer and fluid flow in a three-dimensional wavy-fin and tube heat exchanger. Int J Heat Mass Transf 40(16):3981–3990

Jang Y-J, Chen H-C, Han J-C (2001) Computation of flow and heat transfer in two-pass channels with 60 deg ribs. J Heat Transf 123(3):563–575

Jones TV, Russell CMB (1980) Heat transfer distribution on annular fins. ASME Paper 78-HT-30

Jung GH, Jung SH (1999) Heat exchanger fin having an increasing concentration of slits from an upstream to a downstream side of the fin. U.S. patent 5,934,363

Kandlikar SG (1987) A general correlation for saturated two-phase flow boiling heat transfer inside horizontal and vertical tubes. In: Ragi EG, Rudy TM, Rabas TJ, Robertson JM (eds) Boiling and condensation in heat transfer equipment, HTD, vol 85, pp 9–20

Kearney SP, Jacobi AM (1996) Local convective behavior and fin efficiency in shallow banks of in-line and staggered, annularly finned tubes. J Heat Transf 118(2):317–326

Kim JH, Jensen M, Jansen K (2002) Fin shape effects in turbulent heat transfer in tubes with helical fins. In: Heat transfer 2002. Proceedings of the 12th international heat transfer conference, vol 4, pp 183–188

Kruckels SW, Kottke V (1970) Investigation of the distribution of heat transfer on fins and finned tube models. Chem Eng Tech 42:355–362

Lehtinen A (2005) Analytical treatment of heat sinks cooled by forced convection. PhD thesis, Tampere university of technology, Tampere, Finland

Leu J-S, Liu M-S, Liaw J-S, Wang C-C (2001) A numerical investigation of louvered fin-and-tube heat exchangers having circular and oval tube configurations. Int J Heat Mass Transf 44:4235–4243

Lindstedt M, Karvinen R (2012) Optimization of isothermal plate fin arrays with laminar forced convection. J Enhanc Heat Transf 19(6):535–547

Manglik RM, Zhang JH, Muley A (2005) Low Reynolds number forced convection in three-dimensional wavy-plate-fin compact channels: fin density effects. Int J Heat Mass Transf 48:1439–1449

McQuiston FC (1978) Correlation of heat, mass, and momentum transport coefficients for plate-fin-tube heat transfer for surfaces with staggered tube. ASHRAE Trans 54(Part 1):294–309

Min J, Webb RL (2000) Condensate carryover phenomena in dehumidifying, finned-tube heat exchangers. Exp Therm Fluid Sci 22:175–182

Min J, Webb RL (2004) Numerical analyses of effects of tube shape on performance of a finned tube heat exchanger. J Enhanc Heat Transf 11:61–74

Min J, Webb RL, Bemisderfer CH (2000) Long-term hydraulic performance of dehumidifying heat-exchangers with and without hydrophilic coatings. Int J HVAC&R Res 6(3):257–272

Mirth DR, Ramadhyani S (1993) Comparison of methods of modeling the air-side heat and mass transfer in chilled water cooling coils. ASHRAE Trans 99(Pt. 2):285–299

Mori Y, Nakayama W (1980) Recent advances in compact heat exchangers in Japan. In: Shah RK, McDonald CF, Howard CP (eds) Compact heat exchangers—history, technology, manufacturing technologies. ASME Symp. HTD, vol 10, pp 5–16

Muzychka YS (2005) Constructal design of forced convection cooled microchannel heat sinks and heat exchangers. Int J Heat Mass Transf 48:3119–3127

Neal SBHC, Hitchcock JA (1966) A study of the heat transfer processes in banks of finned tubes in cross flow, using a large scale model technique. In: Proceedings of the third international heat transfer conference, vol 3. American Institute of Chemical Engineers, pp 290–298

O'Brien JE, Sohal MS, Wallstedt PC (2001) Local heat transfer and pressure drop for finned-tube heat exchangers using oval tubes and vortex generators. In: Proceedings of 2001 ASME international mechanical engineering congress and exposition, ASME, New York, Paper No. IMECE2001/HTD-24118

O'Connor JM, Pasternak SF (1976) Method of making a heat exchanger. U. S. Patent 3,947,941

Onishi H, Inaoka K, Matsubara K, Suzuki K (1999) Numerical analysis of flow and conjugate heat transfer of two-row plate-finned tube heat exchanger. In: Shah RK, Bell KJ, Honda H, Thonon B (eds) Proceedings of the international conference on compact heat exchangers and enhancement technology for the process industries. Begell House Inc., New York, pp 175–183

Rabas TI, Huber FV (1989) Row number effects on the heat transfer performance of in-line finned tube banks. Heat Transf Eng 10(4):19–29

Rabas TJ, Myers GA, Eckels PW (1986) Comparison of the thermal performance of serrated high-finned tubes used in heat-recovery systems. In: Chiou JP, Sengupta S (eds) Heat transfer in waste heat recovery and heat rejection systems. ASME Symp. HTD, vol 59, pp 33–40

Saboya FEM, Sparrow EM (1974) Local and average heat transfer coefficients for one-row plate fin and tube heat exchanger configurations. J Heat Transf 96:265–272

Saboya FEM, Sparrow EM (1976a) Experiments on a three-row fin and tube heat exchanger. J Heat Transf 98:520–522

Saboya FEM, Sparrow EM (1976b) Transfer characteristics of two-row plate fin and tube heat exchanger configurations. Int Heat Mass Transf 19:41–49

Saha A (2008) Effect of the number of periodic module on flow and heat transfer in a periodic array of cubic pin-fins inside a channel. J Enhanc Heat Transf 15(3):243–260

Sheui TWH, Tsai SF, Chiang TP (1999) Numerical study of heat transfer in two-row heat exchangers having extended fin surfaces. Numer Heat Transf Part A 35(7):797–814

Somchai W, Yutasak C (2005) Effect of fin pitch and number of tube rows on the air side performance of herringbone wavy fin and tube heat exchangers. Energy Convers Manag 46:2216–2231

Tanaka T, Hatada T, Itoh M, Senshu T, Katsumata N, Michizuki Y, Terada H, Izushi M, Sato M, Tsuji H, Nagai M (1994) Fin-tube heat exchanger. U.S. patent 5,360,060

Tao YB, He YL, Wu ZG, Tao WQ (2007) Three-dimensional numerical study and field synergy principle analysis of wavy fin heat exchangers with elliptic tubes. Int J Heat Fluid Flow 28 (6):1531–1544

Teertstra P, Yovanovich MM, Culham JR (2000) Analytical forced convection modeling of plate fin heat sinks. J Electron Manuf 10(4):253–261

Torikoshi K, Xi G (1995) A numerical study of flow and thermal fields in finned tube heat exchangers. In: Proceedings of the IMECE, HTD, vol 317-1, pp 453–458

Torikoshi K, Xi GN, Nakazawa Y, Asano H (1994) Flow and heat transfer performance of a plate fin-and-tube heat exchanger (1st report: effect of fin pitch). In: Heat transfer 1994. Proceedings of the 10th international heat transfer conference, vol 4, pp 411–416

Tsai SF, Sheu TWH, Lee SM (1999) Heat transfer in a conjugate heat exchanger with a wavy fin surface. Int J Heat Mass Transf 42:1735–1745

Ueda H, Hatada T, Kunugi N, Ooucgi T, Sugimoto S, Shimizu T, Kohno K (1994) Heat transfer fins and heat exchanger. U.S. patent 5,353,886

Valencia A, Fiebig M, Mitra NK (1996) Heat transfer enhancement by longitudinal vortices in a fin-tube heat exchanger element with flat tubes. J Heat Transf 118(1):209–211

Wang CC, Fu WL, Chang CT (1997) Heat transfer and friction characteristics of typical wavy fin-and-tube heat exchangers. Heat Transf Friction Charact 14:174–186

Wang CC, Jang JY, Chiou NF (1999) A heat transfer and friction correlation for wavy fin-and-tube heat exchangers. Int J Heat Mass Transf 42:1919–1924

Wang C-C (2000) Recent progress on the air-side performance of fin-and-tube heat exchangers. Int J Heat Exchanges 1:49–76

Wang C-C, Chi K-Y, Chang C-J (2000) Heat transfer and friction characteristics of plain fin-and-tube heat exchangers, part II: correlation. Int J Heat Mass Transf 43:2693–2700

Webb RL (1990) The flow structure in the louvered fin exchanger geometry. SAE Int. Congress and Exposition, SAE 900722

Webb RL, Gupte N (1990) Design of light weight heat exchangers for air-to-two phase service. In: Shah RK, Kraus AD, Metzger D (eds) Compact heat exchangers: a Festschrift for A. L. London. Hemisphere Publishing Corp., Washington, pp 311–334

Webb RL, Iyengar A (2001) Oval finned tube condenser and design pressure limits. J Enhanc Heat Transf 8:147–158

Webb RL, Kim NY (2005) Principles of enhanced heat transfer. Taylor & Francis, New York

Wieting AR (1975) Empirical correlations for heat transfer and flow friction characteristics of rectangular offset fin heat exchangers. J Heat Transf 97:488–490

Wu JM, Tao WQ (2007) Numerical computation of laminar natural convection heat transfer around a horizontal compound tube with external longitudinal fins. Heat Transf Eng 28(2):93–102

Xin RC, Li HZ, Kang HJ, Li W, Tao WQ (1994) An experimental investigation on heat transfer and pressure drop characteristics of triangular wavy fin-and-tube heat exchanger surfaces. J Xi'an Jiaotong Univ 28(2):77–83

Yilmaz A, Büyükalaca O, Yilmaz T (2000) Optimum shape and dimensions of ducts for convective heat transfer in laminar flow at constant wall temperature. Int J Heat Mass Transf 43:767–775

Youn B, Kim YS (1998) Heat exchanger fins of an air conditioner. U.S. patent 5725625

Yun J-Y, Kim H-Y (1997) Structure of heat exchanger. U.S. patent 5697432

Chapter 6
Internally Finned Tubes and Spirally Fluted Tubes

Figures 1.4 and 6.1 show the internally finned tubes with dimensions. The internally finned tubes may be used for both laminar and turbulent flows. Another dimension for helical fins is the helix angle, α. There is no flow separation on the internal fin.

Watkinson et al. (1975a) reported Nu and f data ($50 < Re_d < 3000$) for steam heating of oil ($180 < Pr < 350$) in 18 different internally finned tubes for the fully developed laminar flow. Marner and Bergles (1978, 1985, 1989), Rustum and Soliman (1988a, b), Hu and Chang (1973), Nandakumar and Masliyah (1975), Soliman and Feingold (1977), Patankar and Chai (1991), Soliman et al. (1980), Soliman and Feingold (1977), Soliman (1979), Prakash and Patankar (1981), Prakash and Liu (1985), Choudhury and Patankar (1985), Shome and Jensen (1996a, b), Soliman (1979), Watkinson et al. (1975a), Bergles and Joshi (1983), Rabas and Mitchell (2000), Zhang and Ebadian (1992a), Al-Fahed et al. (1998), Shome (1998) and Kelkar and Patankar (1990) have worked experimentally or numerically or both on the laminar flow through internally finned tubes (Figs. 6.2, 6.3, 6.4, 6.5, 6.6 and 6.7, Tables 6.1, 6.2, 6.3, 6.4 and 6.5). The flow was laminar, either fully developed or entrance region developing flow with free convection effects. Microfin tubes and segmented internally finned tubes were used with a wide Prandtl number range ($24 < Pr < 8130$).

Wang et al. (2009) numerically studied the performance of internally finned tubes. They compared the performance of S-shaped, Z-shaped and V-shaped lateral fin profiles using realizable k-ε turbulence model. The cross-sectional view of S-shaped, Z-shaped and V-shaped fin profiles has been shown in Fig. 6.8. They calculated Nu and f for blocked core tube with internal longitudinal plain fins having lateral fin profiles. They have also presented suitable correlations. The performance of three lateral fin profiles has been compared under three different constraints: identical mass flow rate, identical pumping power and identical pressure drop. Figure 6.9 illustrates the variation of Nusselt number and friction factor with Reynolds number for all the three lateral fin profiles. It has been reported that for $Re < 7000$, the maximum friction factor was noted for the V-shaped fin.

© The Author(s), under exclusive license to Springer Nature Switzerland AG 2020
S. K. Saha et al., *Heat Transfer Enhancement in Externally Finned Tubes and Internally Finned Tubes and Annuli*, SpringerBriefs in Applied Sciences and Technology, https://doi.org/10.1007/978-3-030-20748-9_6

Fig. 6.1 Definition of
dimensions of the internally
finned tube (Webb and Kim
2005)

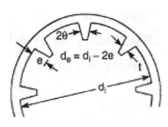

Fig. 6.2 *Nu* vs. *X*+ for
constant wall temperature
and constant heat flux
boundary conditions
(Rustum and Soliman
1988a)

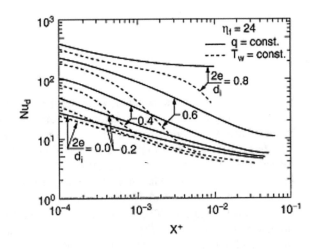

Fig. 6.3 Influence of
Rayleigh number on *Nu* for
laminar entrance region flow
with electric heat input
($d_i = 13.9$ mm, $\eta_f = 10$,
$e/d_i = 0.11$) (Rustum and
Soliman 1990)

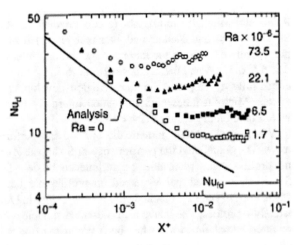

On the other hand, for $Re > 7000$, S-shaped finned tubes showed maximum
pressure drop characteristics. The Nusselt number for tube with Z-shaped fin and
V-shaped fin showed the maximum and minimum, respectively. The velocity and
temperature profiles for S-shaped and Z-shaped fin profiles have been observed to be

Fig. 6.4 Effect of natural convection on f_{Dh}, Re_{Dh} and Nu_{Dh} for an internally finned tube having 11 fins, for constant heat flux (Zhang and Ebadian 1992b)

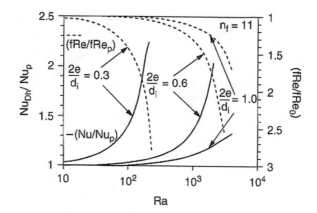

Fig. 6.5 Effect of number of fins on the laminar flow Nusselt number at a low heat flux ($3 \times 10^5 \leq Ra \leq 2 \times 10^6$); (**a**) internal finned tubes with $e/d_i = 0.05$ and $\alpha = 30°$, (**b**) microfin tubes with $e/d_i = 0.015$ and $\alpha = 30°$ (Shome and Jensen 1996a)

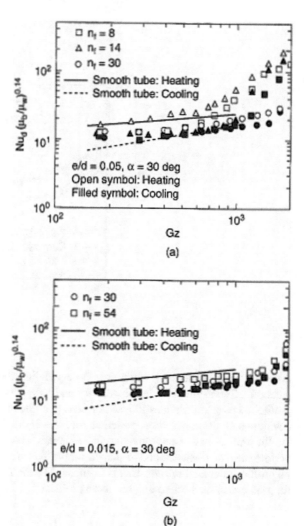

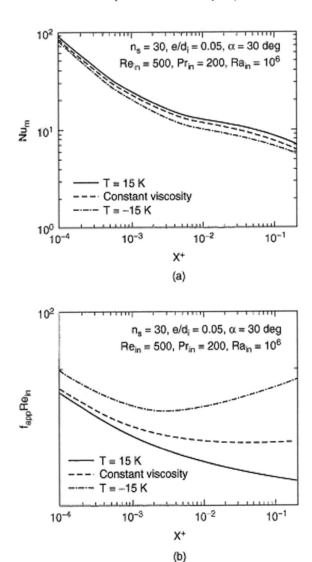

Fig. 6.6 Numerical results on the effect of variable viscosity for an internal fin tube with $\eta_f = 30$, $e/d_i = 0.05$, and $\alpha = 30°$. Calculations were done at $Re_{in} = 500$, $Pr_{in} = 200$ and $Ra_{in} = 10^6$. (**a**) Average Nusselt number, (**b**) friction factor (Shome and Jensen, 1996b)

symmetric and parabolic, while for V-shaped fins they are unsymmetrical and parabolic. The asymmetry in the velocity and temperature profiles of V-shaped fin profile is because it divides the section into two unequal parts. The fitness-relative deviations for the correlations presented for Nusselt number and friction factor for all the fin profiles have been tabulated in Table 6.6. They concluded that the thermal performance of S-shaped and Z-shaped fin profiles are better than that of V-shaped fin profile under all three constraints assumed. The thermal performance of Z-shaped fin profile was the best among the three profiles.

Fig. 6.7 Illustration of in-line and staggered segmented internal fins analysed by Kelkar and Patankar (1990)

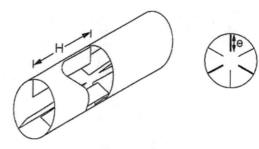

Staggered fins

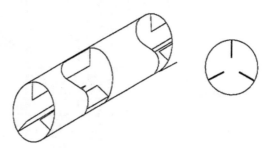

Inline fins

Table 6.1 Enhancement ratios provided by internally finned tubes for fully developed laminar flow (Webb and Kim 2005)

n_f	e/d_i	A/A_p	f/f_p	$\theta k_f/k$			
				$(Nu_d/Nu_p)\tau$		$(Nu_d/Nu_p)_{HI}$	
				5	∞	5	∞
4	0.1	1.26	1.24	1.04	1.04	1.05	1.07
	0.2	1.51	1.91	1.28	1.30	1.38	1.45
	0.3	1.76	3.28	2.25	2.44	2.47	2.84
	0.4	2.02	4.80	3.73	4.40	3.61	4.52
8	0.1	1.51	1.57	1.06	1.06	1.08	1.10
	0.2	2.02	3.53	1.29	1.31	1.50	1.56
	0.3	2.58	8.67	2.41	2.50	3.79	4.84
	0.4	3.07	14.5	8.07	9.64	7.81	10.45
16	0.1	2.02	2.02	1.03	1.03	1.06	1.06
	0.2	3.04	5.93	1.09	1.10	1.18	1.21
	0.3	4.06	22.2	1.45	1.47	2.07	2.18
	0.4	5.07	60.8	8.59	8.66	17.4	24.4
24	0.1	2.53	2.26	1.01	1.01	1.02	1.02
	0.2	4.06	6.99	1.02	1.03	1.05	1.06
	0.3	5.58	31.7	1.13	1.13	1.29	1.32
	0.4	7.11	172.0	3.29	3.30	9.31	10.5
32	0.1	3.04	2.36	1.00	1.00	1.00	1.01
	0.2	5.08	6.99	1.00	1.00	1.01	1.02
	0.3	7.11	36.3	1.03	1.03	1.08	1.09
	0.4	9.15	355.0	1.66	1.66	2.81	2.91

Table 6.2 Values of $\theta k_t/k$ for different fluid–material combinations at 25 °C

Fluid	Pr	Copper	Aluminium	Steel
Air	0.7	16,000	8000	2000
Water	6.0	670	330	80
Oil	1200	2700	1300	400

Table 6.3 Thermal entrance lengths for internally finned tubes (Webb and Kim 2005)

e/d_i	n_f 0	4	8	16	24
Heal flux boundary condition					
0.0	0.0442				
0.1		0.0451	0.0462	0.0475	0.0466
0.2		0.0458	0.0524	0.0544	0.0518
0.3		0.0287	0.0412	0.0596	0.0630
0.4		0.0109	0.0589	0.0028	0.0102
Constant wall temperature boundary condition					
0.0	0.0357				
0.1		0.0259	0.0209	0.0157	0.0131
0.2		0.0255	0.0197	0.0128	0.0107
0.3		0.0274	0.0234	0.0093	0.0064
0.4		0.0088	0.0055	0.0296	0.0093

Table 6.4 Hydrodynamic entrance length parameters for internally finned tubes (Webb and Kim 2005)

e/d_i	n_f 0	8	16	24
Hydrodynamic entrance length				
0.00	0.0415			
0.15		0.0433	0.0438	0.0417
0.30		0.0320	0.0540	0.0622
0.50		0.0052	0.0024	0.0014
Incremental pressure drop, $K(\infty)$				
0.00	1.25			
0.15		2.44	4.11	5.40
0.30		2.85	10.70	23.50
0.50		1.58	1.79	1.93

Table 6.5 Results of segmented and continuous internal fins for $e/d_i = 0.15$

Geometry	fRe/fRe_p	Nu_d/Nu_p
Continuous fins	2.17	1.18
Inline segmented fins	1.71	1.25
Staggered segmented fins	2.19	1.12

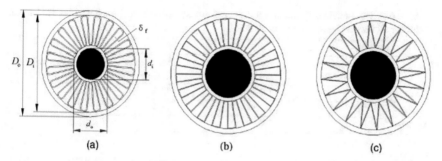

Fig. 6.8 Cross-sectional view of (**a**) S-shaped, (**b**) Z-shaped and (**c**) V-shaped fin profiles (Wang et al. 2009)

Fig. 6.9 Variation of Nusselt number and friction factor with Reynolds number for all three lateral fin profiles. (**a**) f vs. Re. (**b**) Nu vs. Re (Wang et al. 2009)

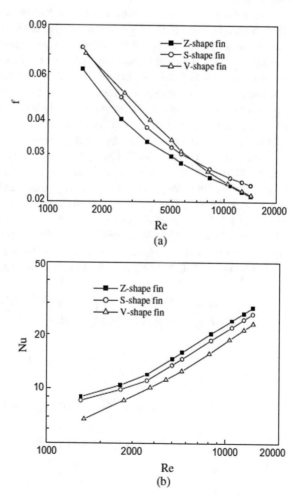

Table 6.6 Fitness-relative deviations for the correlations presented for Nusselt number and friction factor for all the fin profiles (Wang et al. 2009)

Fin	Nu		f	
	Max relative deviation (%)	Average relative deviation (%)	Max relative deviation (%)	Average relative deviation (%)
S-shape	19.9	1.4	22.3	2.7
Z-shape	17	1	18.3	2.1
V-shape	11.8	1.9	13	1.7

Fabbri (1998, 1999), Zeitoun and Hegazy (2004), Olson (1992), Alam and Ghoshdastidar (2002), Saad et al. (1997), Kumar (1997), Yu et al. (1999), Liu and Jensen (1999), Sarkhi and Nada (2005), Wang et al. (2008a, b), Eckert and Irvine (1960), Yu and Tao (2004), Shih et al. (1995) and Park and Ligrani (2005) have carried out similar works.

Liu and Jensen (2001) investigated the performance of internally finned tubes. They presented the effect of geometrical parameters of the fin and tube on heat transfer and pressure drop characteristics. They reported that when the region between the fins is considerable, Nu and f were increased with increase in helix angle of the fin and the number of total fins. For large inter-fin spacing, they observed that the performance of rectangular and triangular fin profiles was similar and superior to that of round fin profile. This has been attributed to the strong interaction between the sharp corners of the rectangular fin tip on the windward side and more turbulence created by the fluid flow.

Hilding and Coogan (1964), Carnavos (1979, 1980), Webb and Scott (1980), Trupp and Haine (1989), Watkinson et al. (1973, 1975b), Kim and Webb (1993), Gowen and Smith (1968), Trupp et al. (1981), Jensen and Vlakancic (1999), Shome and Jensen (1996a), El-Sayed et al. (1997), Braga and Saboya (1986), Said and Trupp (1984), Patankar et al. (1979), Ivanović et al. (1990), Liu and Jensen (1999, 2001), Kim et al. (2002), Wolfstein (1988), Bhatia and Webb (2001) and Webb (1981) have worked with turbulent flow conventional internally finned tubes with $10,000 \leq Re_{Dh} \leq 150,000$ (Figs. 6.10, 6.11, 6.12, 6.13 and 6.14, Table 6.7).

$$\frac{Nu_{Dh}}{Nu_p} = \frac{hD_h/k}{h_p d_i/k} = \left[\frac{d_i}{d_{im}}\left(1 - \frac{2e}{d_i}\right)\right]^{-0.2} \left(\frac{d_i D_h}{d_{im}^2}\right)^{0.5} \sec^2\alpha \tag{6.1}$$

$$\frac{f_{Dh}}{f_p} = \frac{d_{im}}{d_i} \sec^{0.75}\alpha \tag{6.2}$$

$$Nu_p = \frac{h_p d_i}{k} = 0.023 Re_p^{0.8} Pr^{0.4} \tag{6.3}$$

$$f_p = 0.046 Re_p^{-0.2} \tag{6.4}$$

Fig. 6.10 Friction data of
Trnpp and Haine (1989) for
internally finned tubes
(Trupp and Haine 1989)

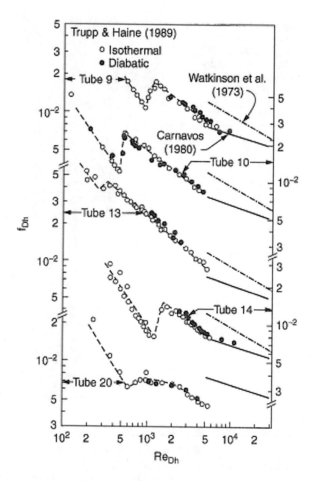

$$\frac{Nu_d}{Nu_p} = \frac{Nu_{Dh}}{Nu_p} \cdot \frac{d_i}{D_h}\left(1 + \frac{2 n_f e}{\pi d_i}\right) \qquad (6.5)$$

$$\frac{f_d}{f_p} = \frac{f_{Dh}}{f_p} \cdot \frac{d_i}{D_h} \qquad (6.6)$$

These works have significantly advanced computational abilities and generated experimental data for three-dimensional internally finned tubes; information on local flow structure measurements, microfin tubes, segmented internally finned tubes and numerical approach on turbulent flow through internally finned tubes may be obtained from the above-mentioned works.

Peng et al. (2016) presented a novel wavy-fin array for heat transfer enhancement in an internally finned tube. They studied the impact of variation of fin geometry on the heat transfer and pressure drop characteristics. Both the experimental and numerical analyses have been carried out. The air-side heat transfer performance

Fig. 6.11 Fully developed
Nusselt number data of
Trupp and Haine (1989) for
Table 8.6 of internally
finned tubes (Trupp and
Haine 1989)

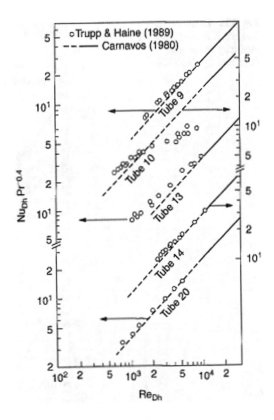

of new fin array has been evaluated in the Reynolds number 2000–20,000. The
geometry of wavy fin arrays has been shown in Fig. 6.15. The fin array was welded
on the inner surface of the internally finned tube. The important geometrical parameters of the fin are wave height (H), space between fins (P_m), width of the wave
(W_m), thickness of the wave (δ_m), wave effective length (L_m) and the wave angle (γ).
The wave angle can be defined as the angle between wave axis and wave edge. The
inner diameter (D_i), inscribed circle diameter (d_{ins}), tube wall thickness (δ_T), fin
height (H_f), fin pitch (W_{1f} and W_{2f}), fin thickness (δ_f) and number of fins (N_t) are the
other important geometrical parameters related to the tube and fin geometries.

The internally wavy finned tube configurations have been shown in Fig. 6.16.
They observed that vortex generation around the wave fin corrugation induces
secondary flow and augments the intensity of turbulence. Further, the velocity
gradient increases reducing the thermal boundary layer, and thus, the heat transfer
augmentation is achieved. An increase in Nusselt number with increase in wave
width and height was reported. Also, with increase in space between waves, the
Nusselt number was observed to decrease. The friction factor on the other hand
increased with wave space and decreased with wave width and wave height. The
corrugation having smaller wave angle resulted in high overall thermal performance.
They concluded that the WP-WP arrangement showed superior performance than
that of WP-WV arrangement for the wave arrays.

Fig. 6.12 Experimental measurements of Trupp et al. (1981): $e/d_i = 0.33$, $t/e = 0.13$, $Re_d = 71,000$. (a) Measured velocity profile, (b) surface shear stress distribution (Trupp et al. 1981)

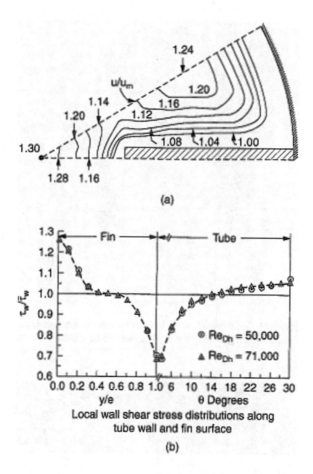

(a)

(b)

Local wall shear stress distributions along tube wall and fin surface

Wang et al. (2014, 2015), Martinez et al. (2015), Luo et al. (2014), Zhang et al. (2013), Promvonge (2015), Lin et al. (2014), Fabbri (1998, 1999), Yu and Tao (2004), Zeitoun and Hegazy (2004), Wang et al. (2008a, b, 2009, 2013), Fabbri (2004, 2005), Dagtekin et al. (2005), Sarkhi and Nada (2005), Islam and Mozumder (2009), García et al. (2012), Rout et al. (2012), Iqbal et al. (2013), Hatami et al. (2014, 2015), Syed et al. (2015), Liu et al. (2013a, b, 2015), Peng and Ling (2011) and Peng et al. (2014) are relevant works on internally finned tubes.

Kim (2015a) investigated 7 three-dimensional dimpled tubes for the heat transfer and pressure drop analysis. Higher heat transfer was obtained for three-dimensional roughness in comparison to two-dimensional roughness. Webb and Kim (2005), Kim and Webb (1989), Nikuradse (1922), Cope (1945) and Dipprey and Sabersky (1963) used the roughened tube for heat transfer and friction factor analyses. Liao et al. (2000) used the tubes consisting of three-dimensional integral roughness with higher roughness heights and compared with tested results of Takahashi et al. (1988), Kuwahara et al. (1989), Wang et al. (2010), Chen et al. (2001), Nivesrangsan et al. (2010), Mahmood and Ligrani (2002) and Nishida et al. (2012) worked with

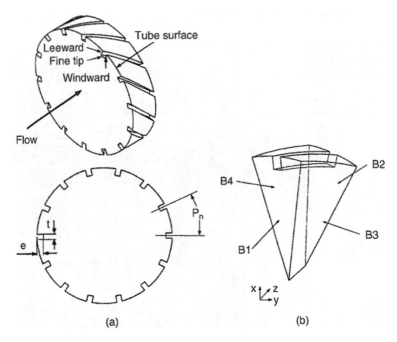

Fig. 6.13 Solid model of the internal fin tube by Liu and Jensen (1999). (**a**) Basic geometry, (**b**) computational geometry model (Liu and Jensen 1999)

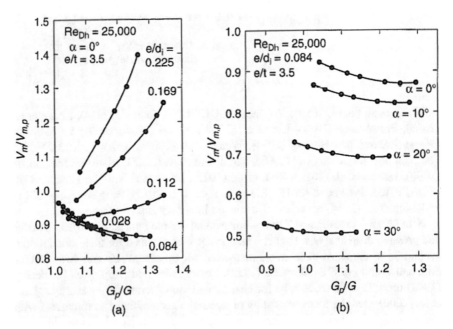

Fig. 6.14 Performance comparison of internally finned tubes and plain tube ($d = 17\ 0.78$ mm) for case VG-2 of Table 6.7 (**a**) effect of e and η_f for $e/t = 3.5$ and $\alpha = 0$, (**b**) effect of η_f and α for $e/d_i = 0.084$ and $e/t = 3.5$ mm. The points on each curve define η_f (from the left, $\eta_f = 5, 8, 12, 16, 25, 32$ and 40 fins) (Webb and Scott 1980)

Table 6.7 Tube geometries (Trupp and Haine 1989)

No.	9	10	13	14	20
d_o (mm)	12.7	9.53	9.53	15.9	12.7
d_i (mm)	10.3	8.00	7.04	13.9	10.4
e (mm)	1.28	1.27	2.29	1.50	1.47
n_f	10	16	10	10	16
α (deg)	0	0	0	0	2.5

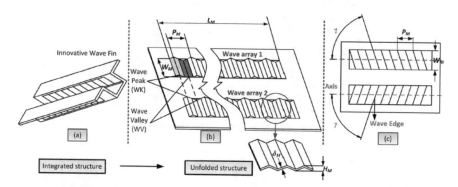

Fig. 6.15 Geometry of wavy fin arrays (Peng et al. 2016)

dimple tubes. Other interesting works of Liao and Xin (1995, 2000), Thianpong et al. (2009), Suresh et al. (2001) and Kumbhar and Sane (2015) were related to three-dimensional roughness in both laminar and turbulent flow conditions.

The objective of Kim (2015a) was to measure the effect of roughness parameters on thermo-hydraulic characteristics. He used 22.2 mm outer diameter and 19.9 mm internal diameter tube and tested the 7 three-dimensional dimpled tubes with geometrical parameter ($0.020 \leq e/D \leq 0.030$, $5.0 \leq P/e \leq 10.0$ and $6.0 \leq Z/e \leq 14.0$). He presented Fig. 6.17 which is for finning disc and sample tube. The testing section geometrical parameters have been detailed in Table 6.8. He experimented and found that optimum dimple height was 0.5 mm, optimum axial pitch was 3.0 mm and optimum circumferential dimple pitch was 5.0 mm. Also, he observed that dimpled tube overcomes the two-dimensional spiral rib tube and three-dimensional diamond-shaped roughness.

Kim (2015b) experimentally investigated the thermo-hydraulic performance of R-410A in an internally flattened microfin tube. The heat transfer enhancement is due to increase in heat transfer surface area and turbulence induced by the microfins. Webb and Kim (2005), Bogart and Thors (1999) and Bergles and Manglik (2013) used typical microfin tubes. Wilson et al. (2003), Webb and Iyengar (2001) and Kim and Kim (2010) experimented and concluded that pressure drop was less in oval-shaped tube compared to round tubes. Webb and Kim (2005), Collier and Thome (1994) and Ghiaasiaan (2008) investigated with rounded tubes, whereas Kim et al. (2002) used oval microfin tubes for investigating evaporation phenomenon. Quiben et al. (2009a, b), Nasr et al. (2010) and Kim et al. (2013) worked with smooth flat tubes for studying evaporation.

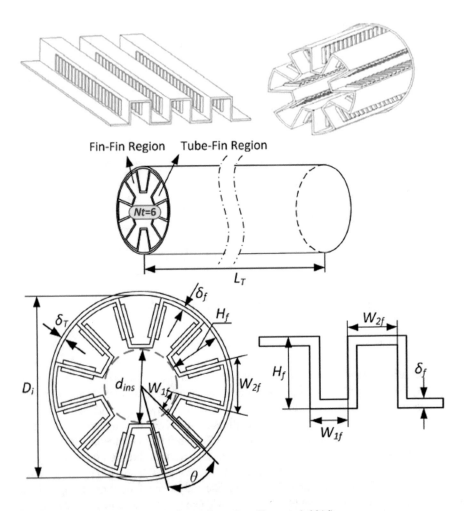

Fig. 6.16 Internally wavy finned tube configurations (Peng et al. 2016)

Fig. 6.17 Finning disc and sample tube (Kim 2015a, b)

Table 6.8 Geometric details of the dimpled tubes

Tube	D	e	z	p	e/D	z/e	p/e	a	b
Smooth	19.9								
e05z5p3	19.9	0.5	5.0	3.0	0.025	1.0	6.0	2.29	1.70
e05z5p5	19.9	0.5	5.0	5.0	0.025	1.0	10.0	2.29	1.70
e05z5p7	19.9	0.5	5.0	7.0	0.025	14.0	10.0	2.29	1.70
e05z3p3	19.9	0.5	3.0	3.0	0.025	6.0	6.0	2.29	1.70
e05z7p3	19.9	0.5	7.0	3.0	0.025	14.0	6.0	2.29	1.70
e04z5p3	19.9	0.4	5.0	3.0	0.020	12.5	7.5	2.11	1.53
e06z5p3	19.9	0.6	5.0	3.0	0.030	8.3	5.0	2.57	1.93

Table 6.9 Geometric details of the smooth and microfin tubes (Kim 2015a, b)

	Microfin tube			Smooth tube		
	Round	AR = 2	AR = 4	Round	AR = 2	AR = 4
A_c (mm^2)	32.24	27.48	18.43	19.6	18.1	12.3
A_{cm} (mm^2)	32.24	27.48	18.43	–	–	–
A_{cr} (mm^2)	33.78	28.16	18.89	–	–	–
A_{ct} (mm^2)	31.77	26.13	17.53	–	–	–
A_{ia} (m^2)	0.0325	0.0325	0.0325	0.0157	0.0157	0.0157
A_{im} (m^2)	0.0205	–	–	0.0157	0.0157	0.0157
D_m (mm)	6.42	–	–	5.0	–	–
D_h (mm)	4.04	3.44	2.31	5.0	4.4	3.0
D_o (mm)	7.00	–	–	–	–	–
D_r (mm)	6.56	–	–	–	–	–
D_t (mm)	6.36	–	–	–	–	–
D_{hm} (mm)	6.42	5.45	3.65	–	–	–
D_{hr} (mm)	6.56	5.46	3.67	–	–	–
D_{ht} (mm)	6.36	5.23	3.51	7.0	–	–
w (mm)	6.56	7.75	8.79	5.0	6.1	6.9
h (mm)	6.56	4.08	2.25	5.0	3.1	1.7
e	0.1	0.1	0.1	–	–	–
n	65	65	65	–	–	–
t (mm)	0.22	0.22	0.22	1.0	–	–
P_w (mm)	31.98	31.98	31.98	15.7	–	–
P_{wm} (mm)	20.17	20.17	20.17	–	–	–
P_{wr} (mm)	20.61	20.61	20.61	–	–	–
P_{wt} (mm)	19.98	19.98	19.98	–	–	–
β	15	15	15	–	–	–
γ	40	40	40	–	–	–

Kim (2015b) presented Table 6.9 in which the geometrical details of microfin tube and smooth tube are listed. It was found that actual surface area of microfin tube was 59% larger than smooth or melt-down surface area. He presented a cross-sectional view of microfin tube in Figs. 6.18 and 6.19. He experimented at

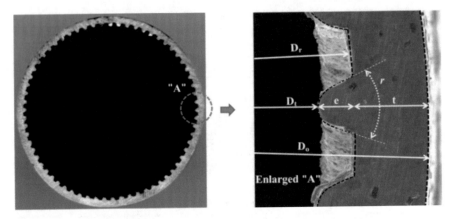

Fig. 6.18 Cross section of microfin tube (Kim 2015a, b)

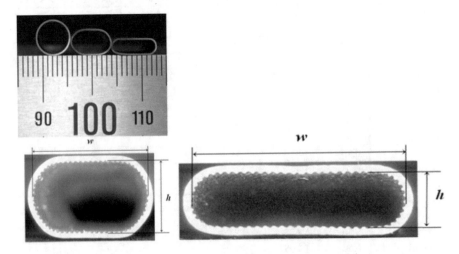

Fig. 6.19 Cross-sectional figures of flat microfin tubes (Kim 2015a, b)

maximum mass flux and quality initially; after gaining stability, quality and mass flux were gradually decreased. He established annulus-side forced convection equations as

$$Nu_{Dh} = 0.0356Re_{Dh}^{0.992}Pr_{w}^{0.3} \quad \text{(round microfin tube)} \tag{6.7}$$

$$Nu_{Dh} = 0.0075Re_{Dh}^{0.993}Pr_{w}^{0.3} \quad (AR = 2 \text{ microfin tube}) \tag{6.8}$$

$$Nu_{Dh} = 0.0024Re_{Dh}^{1.113}Pr_{w}^{0.3} \quad (AR = 4 \text{ microfin tube}) \tag{6.9}$$

and it is valid in $1400 \leq Re_{Dh} \leq 4200$. He proposed the correlation for vapour quality in tube as

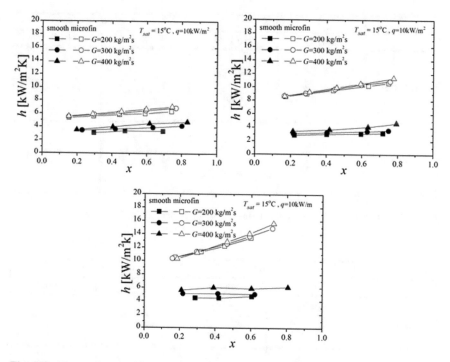

Fig. 6.20 Evaporation heat transfer coefficients in microfin or smooth tubes (Kim 2015a, b)

Table 6.10 Heat transfer enhancement factors and frictional pressure drop penalty factors (Kim 2015a, b)			G (kg/m^{2}s)		
			200	300	400
h/h_s (EF)		Round	1.75	1.65	1.51
		AR = 2	3.81	3.04	2.65
		AR = 4	2.77	2.55	2.23
$(dp/dz)_f/(dp/dz)_{f,s}$ (PF)		Round	0.55	0.88	0.88
		AR = 2	0.63	0.76	0.83
		AR = 4	0.64	0.70	0.68

$$X_{ave} = X_{in} + {}^\Delta x/_2 \qquad (6.10)$$

where Δx is the change of vapour quality. He concluded from the experimental results that heat transfer coefficients of microfin tubes were much higher than that of smooth tubes, and it is shown in Fig. 6.20. He presented Table 6.10 for mass flux and enhancement factor interdependence for all the three tube geometries. The predicted mass flow rate in different geometries was presented in Fig. 6.21, and the effect of quality was presented in Fig. 6.22. He experimentally found that heat transfer coefficient increased as aspect ratio increase and presented it in Fig. 6.23 for both microfin and smooth tubes. He compared his data with other researchers and established Table 6.11 which shows root mean square error relative to Kim

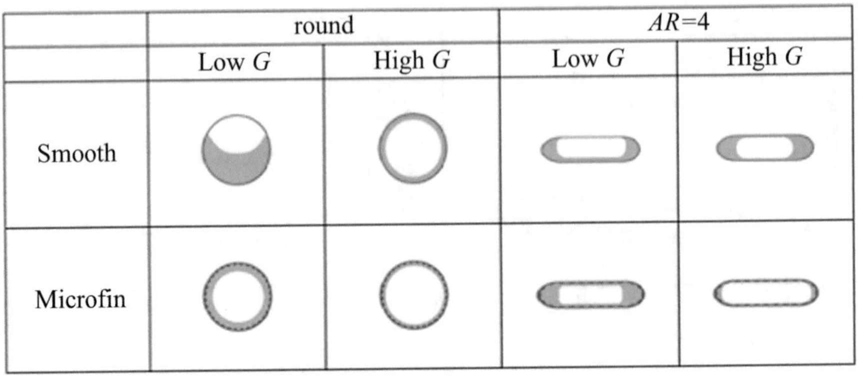

	round		AR=4	
	Low G	High G	Low G	High G
Smooth				
Microfin				

Fig. 6.21 Estimated flow patterns in flat tubes (mass flux effect) (Kim 2015a, b)

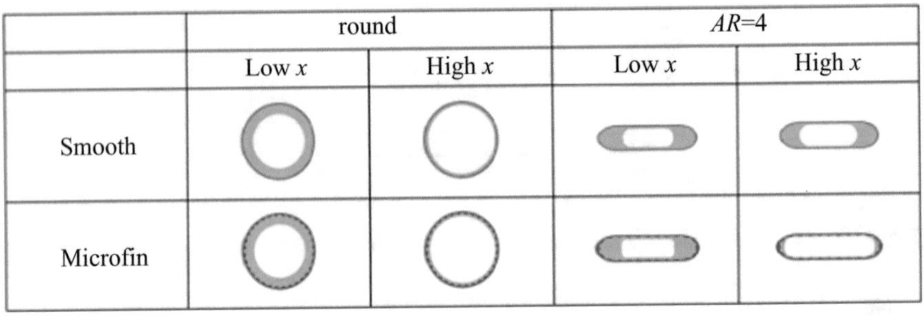

	round		AR=4	
	Low x	High x	Low x	High x
Smooth				
Microfin				

Fig. 6.22 Estimated flow patterns in flat tubes (quality effect) (Kim 2015a, b)

(2015a, b) results. He calculated pressure drop consisting of frictional drop and acceleration pressure drop. The pressure increased with increase in mass flux or quality, and it is shown in Fig. 6.24. Finally, he concluded that heat transfer enhancement ratio increased as mass flux decreased with minimum pressure drop penalty, and it was less than 1.0.

Kim (2016) carried out an analysis to optimize the thermo-hydraulic performance of internally finned tube having variable fin thickness. The thickness variation was in the direction normal to that of the fluid flow. The thermal resistance offered by different internally finned tubes has been shown in Fig. 6.25. The different finned tubes having circular-sector fin, concave fin, convex fin and straight fin have been referred to as case A, case B, case C and case D, respectively. The decrease in thermal resistance with increase in pumping power was noted. The optimization parameters for tubes with straight fins, circular-sector fins and tube having fin with varying thickness have been presented in Table 6.12, and the comparison was made. The optimal fin number variation with the ratio of optimal thermal resistances has been shown in Fig. 6.26. The drop of 12% in thermal resistance has been observed for concave fins over that of straight fins. They reported that the percentage of

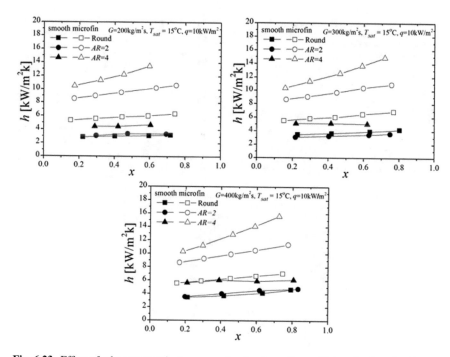

Fig. 6.23 Effect of tube aspect ratio on evaporation heat transfer coefficient for the microfin and smooth tubes (Kim 2015a, b)

Table 6.11 RMS errors of heat transfer coefficient and frictional pressure drops for microfin tubes evaluated by different researchers (Kim 2015a, b)

		RMSE		
	Correlation	Round	AR = 2	AR = 4
h (W/m²K)	Koyama et al. (1995)	0.67	0.22	0.27
	Kido et al. (1995)	0.54	0.62	0.81
	Thome et al. (1997)	0.32	0.65	0.58
	Goto et al. (2001)	0.61	0.26	0.32
	Newell and Shah (2001)	0.08	0.26	0.55
	Yun et al. (2002)	0.38	0.49	0.72
	Cavallini et al. (2006)	0.45	0.18	0.34
	Chamra and Mago (2007)	0.15	0.34	0.61
	Hamilton et al. (2008)	0.45	0.63	0.81
$(dP/dz)/f$ (Pa/m)	Kuo and Wang (1996)	0.30	0.06	0.86
	Cavallini et al. (1997)	0.18	0.36	0.81
	Choi et al. (2001)	0.15	0.32	0.60
	Newell and Shah (2001)	0.37	0.16	0.87
	Goto et al. (2001) (Φ_v)	0.16	0.27	0.84
	Goto et al. (2001) (Φ_l)	0.11	0.42	0.81
	Bandarra Filho et al. (2004)	0.25	0.82	0.75
	Wu et al. (2013)	0.44	0.19	0.90

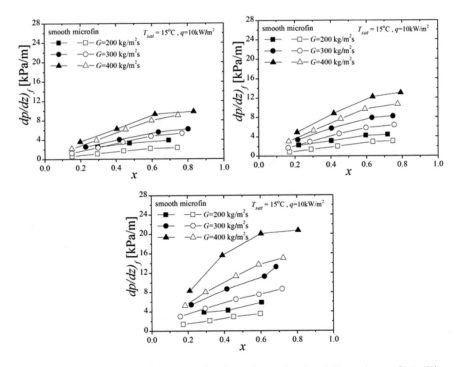

Fig. 6.24 Frictional pressure drops in microfin and smooth tubes (effect of mass flux) (Kim 2015a, b)

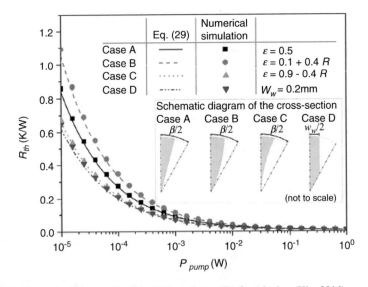

Fig. 6.25 Thermal resistance offered by different internally finned tubes (Kim 2016)

Table 6.12 Optimization parameters for tubes with straight fins, circular-sectored fins and tube having fin with varying thickness (Kim 2016)

	Straight-finned tube	Circular-sectored finned tube	Variable thickness finned tube
	Constraints		
Tube length (L)	10 cm		
Tube radius (r_0)	1 cm		
Pumping power (P_{pump})	0.1 mW		
Solid	Aluminium [$k_t = 175$ W/(m K)]		
Fluid	Water [$k_f = 0.613$ W/(m K), $q = 4179$ J/(kg K), $\rho_f = 997$ (kg/m^2), $\mu_f = 855 \cdot 10^{-4}$ kg/(m s)]		
	Results		
β	0.180	0.196	0.242
Pin number, N	35	32	26
Porosity, z	–	0.845	$1.00 - 0.38R^{2.40}$
Pin thickness, w_w (mm)	0.185	–	–
Hydraulic diameter, D_h (mm)	1.50	1.56	1.88
Nu_{Dh}	1.79	1.81	2.98
$f Re_{Dh}$	10.4	10.6	13.0
Surface area (m^2)	0.0675	0.0682	0.0552
Flow rate (cm^2/s)	5.67	5.97	6.44
$R_{th/rp}$ (K/W)	0.0424	0.0402	0.0372
R_{ch}/rtv (K/W)	0.0292	0.0279	0.0261
R_{ch} (K/W)	0.0716	0.0681	0.0633
Schematic of fin (not to scale)			

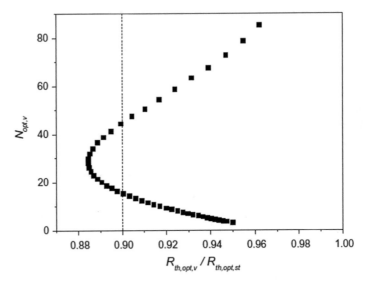

Fig. 6.26 Optimal fin number variation with the ratio of optimal thermal resistances (Kim 2016)

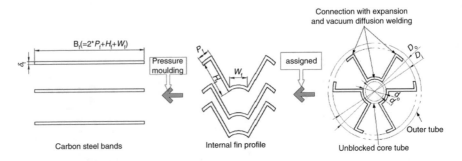

Fig. 6.27 Fin manufacturing procedure (Duan et al. 2018)

reduction in thermal resistances is dependent on the pumping power and the tube length. They concluded that the performance of tube having internal fin with varying thickness was better than that of straight finned tube and proposed that they can be used for thermal systems in cooling equipment.

Bar-Cohen and Kraus (1990), Huq et al. (1998), Hu and Chang (1973), Webb and Scott (1980), Kim et al. (2002, 2010), Bergman et al. (2011) and Kim and Kim (2007) are others who carried out similar works on internally finned tubes.

Blossom-shaped internal fins for a double-tube structured internal fin tube has been used by Duan et al. (2018) in order to study its thermo-hydraulic behaviour. The study has been carried out for the turbulent flow regime. Both the experimental and numerical analyses have been performed. They considered a sample having three blossom-shaped fins. The fin manufacturing procedure has been illustrated in Fig. 6.27. The blossom finned tube has been shown in Fig. 6.28. It consists of an outer tube and an unblocked core tube along with internal fins.

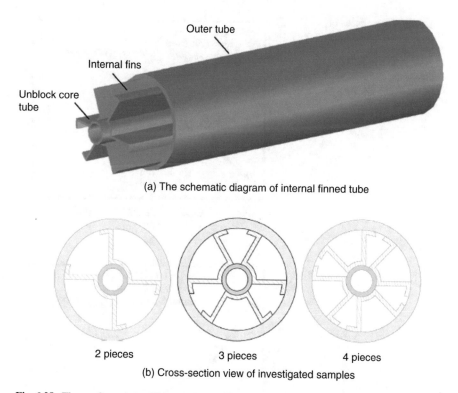

(a) The schematic diagram of internal finned tube

2 pieces   3 pieces   4 pieces

(b) Cross-section view of investigated samples

Fig. 6.28   The configurations of internal finned tube samples. (**a**) The schematic diagram of internal finned tube. (**b**) Cross-section view of investigated samples (Duan et al. 2018)

The variation of Nu and f with Re obtained from experimental data, numerical data and the correlations have been presented in Fig. 6.29 for comparison. They reported that the realizable k-ε turbulence model was better to obtain results in the given Reynolds number range of 3250–19,650. Also, a uniform temperature and velocity field distribution has been observed for increasing number of fins. For cost-effective performance, the optimal d_o/D_i ratio has been proposed to be less than or equal to 0.28. They concluded that the results for thermo-hydraulic performance of blossom-shaped finned tube were inferior to that of wavy finned tube. But, in special applications like exhaust gas heat recovery systems where there is strict restriction on pressure drop, the blossom-shaped fins may be preferable.

Choi et al. (2010), Huang et al. (2014), Song et al. (2010), Ma et al. (2012), Lemouedda et al. (2011), Kim and Webb (1993), Rowley and Patankar (1984) and Liu et al. (2015) also investigated the heat transfer and pressure drop characteristics in internally finned tubes.

Figure 6.30 shows spirally fluted tubes, which are the extended surface obtained by deforming the tube wall to form spiral flutes. Yampolsky (1983) and Marto et al. (1979) worked with spiralled fluted tubes. Panchal and France (1986), Ravigururajan and Bergles (1995), Obot et al. (1991), Barba et al. (1983), Baughn et al. (1993) and

Fig. 6.29 Comparisons of Nusselt number and Darcy fiction factor with Reynolds number among correlation, experimental and numerical data. (**a**) *Nu* vs. *Re*. (**b**) *f* vs. *Re* (Duan et al. 2018)

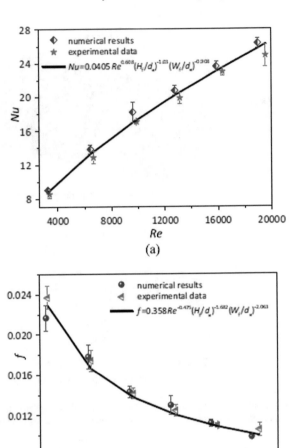

Perera and Baughn (1994) investigated spirally fluted tubes. The local heat transfer coefficients are such as to conclude that thermal development is very rapid for spirally fluted tubes. The heat transfer coefficients on the windward side are higher than those at the leeward side.

Ravigururajan and Bergles (1995), Blumenkrantz and Taborek (1971), Richards et al. (1987), Srinivasan and Christensen (1992), Arnold et al. (1993), Garimella and Christensen (1995a, b) and Srinivasan et al. (1994) worked with spirally indented enhanced tube (Tables 6.13 and 6.14).

Equations (6.11)–(6.16) give the relevant information for in-tube flow and annular flow through spirally indented tubes.

$$\frac{hd_e}{k} = C_i \left(\frac{d_e G}{\mu}\right)^{0.8} Pr^{1/3} \left(\frac{\mu}{\mu_w}\right)^{0.14} \tag{6.11}$$

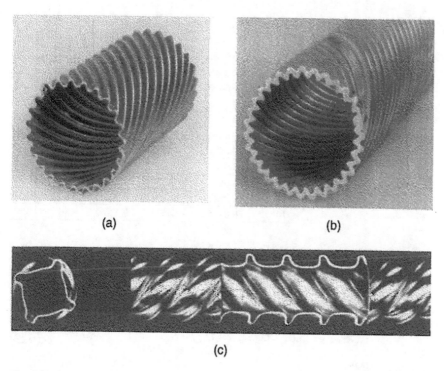

Fig. 6.30 Spirally fluted tubes: (**a**) stainless steel tube, (**b**) tube made of aluminium, (**c**) spirally indented tube (Yampolsky 1983)

Table 6.13 Dimensionless geometric parameters of tubes tested (Richards et al. 1987)

Tube	d/d_c	e/d_c	e/p	e/d_c	p/d_c	A_c/A_e
1	1.56	0.179	0.238	0.278	1.168	0.89
2	1.59	0.186	0.479	0.296	0.618	1.41
3	1.77	0.217	0.416	0.385	0.925	1.48
4	1.23	0.093	0.179	0.114	0.637	0.92
5	1.38	0.139	0.272	0.192	0.706	1.02
6	1.49	0.165	0.515	0.247	0.408	1.11
7	1.93	0.241	0.349	0.465	1.332	1.60
8	1.90	0.237	0.704	0.449	0.638	2.04
9	1.71	0.208	0.275	0.356	1.295	1.22
10	1.83	0.226	0.221	0.414	1.873	1.28
11	2.07	0.258	0.225	0.534	2.373	1.39
12	1.68	0.202	0.500	0.388	0.776	1.36

Table 6.14 Curve fit and
performance parameters
for doubly fluted tubes as
described in Table 6.13
(j/j_p and η at $Re = 10{,}000$)

Tube	C_i	B	n	j/j_p	η
1	0.0442	4.07	0.297	1.84	0.21
2	0.0681	3.79	0.253	2.96	0.24
3	0.0440	6.33	0.276	2.15	0.13
4	0.0455	0.45	0.125	2.45	0.52
5	0.0496	1.37	0.208	2.30	0.34
6	0.0632	0.76	0.117	3.14	0.37
7	0.0596	8.73	0.305	3.02	0.20
8	0.0501	6.81	0.235	2.56	0.12
9	0.0487	4.24	0.244	2.61	0.17
10	0.0480	5.55	0.260	2.25	0.14
11	0.0526	14.73	0.365	2.63	0.30
12	0.0495	3.51	0.220	2.56	0.25

$$f = B \left(\frac{d_e G}{\mu} \right)^n \tag{6.12}$$

$$f = 0.554 \left(\frac{64.0}{Re_{Dvi} - 45.0} \right) \left(\frac{e}{D_{vi}} \right)^{0.384} \left(\frac{p}{D_{vi}} \right)^{(-1.454 + 2.083 e/D_{vi})} \left(\frac{\alpha}{90} \right)^{-2.42} \tag{6.13}$$

$Re_{Dvi} > 1500$

$$f = 1.209 (Re_{Dvi})^{-0.261} \left(\frac{e}{D_{vi}} \right)^{(1.26 - 0.05 p/D_{vi})} \left(\frac{p}{D_{vi}} \right)^{(-1.66 + 2.033 e/D_{vi})} \left(\frac{\alpha}{90} \right)^{(-2.669 + 3.67 e/D_{vi})}$$

$Re_D \leq 5000$

$$\tag{6.14}$$

$$Nu_{Dvi} = 0.014 (Re_{Dvi})^{0.842} \left(\frac{e}{D_{vi}} \right)^{0.067} \left(\frac{p}{D_{vi}} \right)^{0.293} \left(\frac{\alpha}{90} \right)^{-0.705} Pr^{0.4} \tag{6.15}$$

$Re_{Dvi} > 5000$

$$Nu_{Dvi} = 0.064 (Re_{Dvi})^{0.773} \left(\frac{e}{D_{vi}} \right)^{-0.242} \left(\frac{p}{D_{vi}} \right)^{-0.108} \left(\frac{\alpha}{90} \right)^{0.599} Pr^{0.4} \tag{6.16}$$

Ma et al. (2012) examined the friction factor characteristics in an internal helical
finned tube where the flowing fluid was water–ethylene glycol mixture. The internal
helical finned tubes were invented by Fujie et al. (1977) and were used by industries
in early days. It has very good heat transfer characteristics with a little pressure
drop penalty. Many researchers Jensen and Vlakancic (1999), Raj et al. (2015),
Al-Fahed et al. (1998), Afroz and Miyara (2007), Liao and Xin (1995) and

Brognaux et al. (1997) found that helical finned tubes were inefficient in laminar regime as heat transfer efficiency was unchanged.

However, Wang et al. (1996), Copetti et al. (2004), Celen et al. (2013), Siddique and Alhazmy (2008), Li et al. (2007) and Mukkamala and Sundaresan (2009) reported higher friction factor of helical finned tubes than that of plain tubes when Reynolds number reached up to transitional regime. Further, many researchers carried out the investigation in turbulent regime. Wang et al. (1996), Carnavos (1980), Jensen and Vlakancic (1999), Webb et al. (2000), Al-Fahed et al. (1993), Li et al. (2007, 2012), Han and Lee (2005), Aroonrat et al. (2013), Ravigururajan and Bergles (1995) and Zdaniuk et al. (2008) concluded that friction factor follows the same trend as of plain tube after transitional regime.

Ma et al. (2012) used 50% water–ethylene glycol mixture. They used geometric parameters such as fin height (e), tube inside diameter (d_i), helix angle (β) and number of fins (N_s) for characterizing the internal helical finned tube and listed them in Table 6.15. The fins were shaped in triangle and trapezoidal, and this has been shown in Fig. 6.31. They conducted experiment in laminar-turbulent flow regime. They presented a plot in which friction factor of tube was drawn against Reynolds

Table 6.15 Geometric parameters of the tested tubes (Ma et al. 2012)

Tube	l (mm)	L (mm)	d_i (mm)	N_s	p (mm)	e (mm)	β (′)	t_b (mm)	θ (′)
Tube-s	3045	2945	16.34	–	–	–	–	–	–
Tube-1	2945	2945	22.48	60	1.18	0.5	45	0.61	43.1
Tube-2	3045	2945	16.662	38	1.38	0.89	60	0.72	43.8

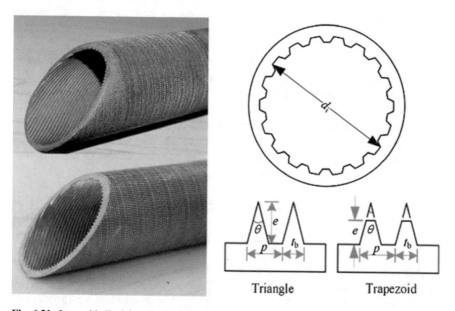

Triangle   Trapezoid

Fig. 6.31 Internal helical finned tube (Ma et al. 2012)

number. They used tube 1, tube 2, smooth tube and results of equation derived by Filonenko. They found that critical Reynolds number was 2160 for tube 1 and 2070 for tube 2. The friction factor follows the same trend of plain tube up to critical Reynolds number. Friction factor increases from Reynolds number 2160 to 2800 for tube 1 and from 2070 to 3137 for tube 2. They observed the entrance region impact on friction factor and concluded that fully turbulent regime begins earlier.

Wang et al. (2017) conducted experiment similar to Ma et al. (2012) for the heat transfer and pressure drop characteristics. They established an equation for the heat transfer coefficient of the internal helically finned tube

$$Nu = (0.0146Re - 8.908)Pr^{0.337} \qquad (6.17)$$

valid for $10,000 < Re < 32,000$ and $17 < Pr < 29$. They concluded that j factor for helically finned tubes was 3.5 times that obtained by plain tube. They calculated efficiency index

$$\eta = \left. \left({}^{j}/_{j_p} \right) \middle/ \left({}^{f}/_{f_p} \right) \right. \qquad (6.18)$$

and found that its value decreased from 1.8 to 1.55 with an increase in Reynolds number in the range of 10,000–32,000.

Bilen et al. (2009) experimentally examined the surface heat transfer and friction factor with different groove shapes. The researchers use turbulence promoters due to significant effectiveness in the enhancement of heat transfer. Webb et al. (1971), Sparrow and Lovell (1980), Kiml et al. (2004), Gee and Webb (1980) and Liu and Jensen (2001) studied roughness-promoted tubes for heat transfer augmentation. Goto et al. (2001, 2003) studied the effects of internally grooved horizontal surface on condensation and evaporation.

Bilen et al. (2009) investigated three types of grooves, namely circular, trapezoidal and rectangular in the range of Reynolds number from 10,000 to 38,000 under uniform wall heat flux boundary condition. Bilen et al. (2009) presented the geometrical shape of all the structures in Fig. 6.32. The experimental setup tube length to diameter ratio was 33. They experimentally found that heat transfer coefficient increased with increase in Reynolds number as shown in Fig. 6.33. It can be understood that fluid mixing by grooves resulted in enhancement of heat transfer. They established the correlation for Nusselt number and friction factor

$$Nu_s = 0.0275Re^{0.781}Pr^{1/3} \qquad (6.19)$$

$$f_s = 1.796Re^{-0.344} \qquad (6.20)$$

Afterwards they concluded the effect of grooved tubes dominated over trapezoidal groove followed by rectangular groove. The maximum heat transfer achieved by circular grooved tube was 63%, followed by trapezoidal grooved tube with 58%, and at last rectangular grooved tube with 47% in comparison to bare smooth tube. It was observed from Fig. 6.34 that maximum Nusselt number was obtained at Reynolds

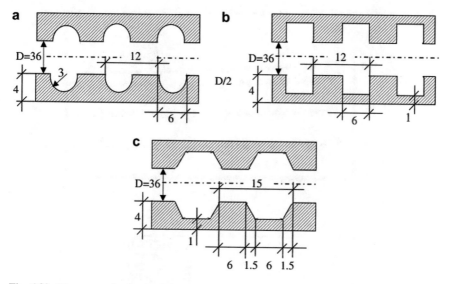

Fig. 6.32 The geometric shapes of the grooved tube, dimensions in mm: (**a**) circular, (**b**) rectangular and (**c**) trapezoidal grooves (Bilen et al. 2009)

number approximately 38,000. The advantage of grooved surface is that it increases turbulence as well as surface area for convective heat transfer. They found that friction factor for all grooved surfaces are comparable and independent of Reynolds number, whereas smooth tube-conjugated friction factor decreased gradually with increased Reynolds number, and it was significantly lower than that of grooved tubes. They developed correlations from experimental data.

For circular grooved tube,

$$Nu = 0.0148 Re^{0.889} Pr^{\frac{1}{3}} \tag{6.21}$$

$$f = 0.356 Re^{.124} \tag{6.22}$$

For rectangular grooved tube,

$$Nu = 0.0339 Re^{0.803} Pr^{\frac{1}{3}} \tag{6.23}$$

$$f = 0.071 Re^{.062} \tag{6.24}$$

For trapezoidal grooved tube,

$$Nu = 0.014 Re^{0.803} Pr^{\frac{1}{3}} \tag{6.25}$$

$$f = 0.0428 Re^{0.107} \tag{6.26}$$

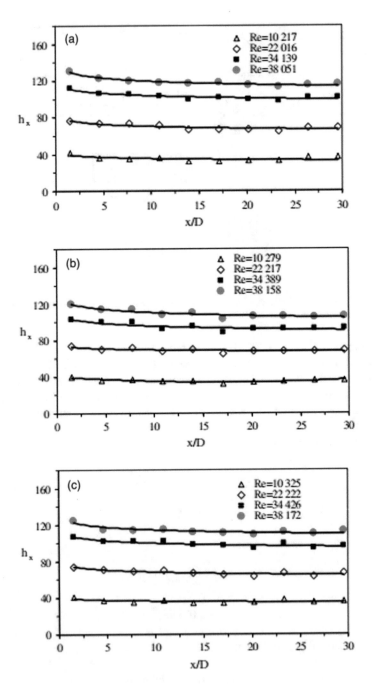

Fig. 6.33 Axial distribution of heat transfer coefficient for (**a**) circular, (**b**) rectangular and (**c**) trapezoidal grooves (Bilen et al. 2009)

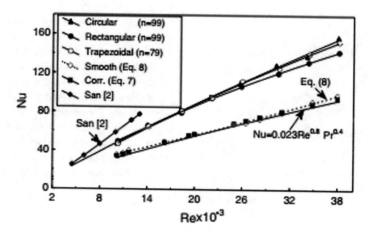

Fig. 6.34 Variation of Nusselt number with Reynolds number for smooth and different grooved tubes (Bilen et al. 2009)

They evaluated their geometries with performance criteria. The heat transfer efficiency was defined as $\eta = (h_a/h_s)_p$ at constant pumping power. They correlated it with all grooved tubes.

$$\eta_{cir} = 1.6356Re^{-0.0261} \tag{6.27}$$

$$\eta_{rec} = 3.054Re^{-0.0939} \tag{6.28}$$

$$\eta_{rec} = 1.4632Re^{-0.0171} \tag{6.29}$$

This has been shown in Fig. 6.35 along with other researchers: Promvonge (2015), Yakut et al. (2004), Promvonge and Eiamsa-Ard (2007), Chang et al. (2007), Manglik and Bergles (1993). The thermal performance (η) was 1.28–1.24 for the circular groove, 1.25–1.22 for the trapezoidal groove and 1.26–1.13 for the rectangular groove for Reynolds number 10,000–38,000. They finally concluded that a 63% increase in heat transfer rate was achieved although it increases manufacturing cost and is advantageous in limiting space heat exchanger.

San and Huang (2006) examined the heat transfer enhancement of transverse ribs containing circular tubes. The objective of the study was to investigate the optimum-height-to-tube-diameter ratio (e/d) and rib-pitch-to-tube-diameter ratio (p/d) for an effective heat transfer enhancement. They presented the dimensions of the nine rib-roughened tubes in Table 6.16; the values of e/d and p/d were also mentioned. They tested and measured Nusselt number of each testing tube and found that Nusselt number linearly varies with varying Reynolds number. They plotted fiction factor coefficients with Reynolds number and found that friction factor for parameters $e/d = 0.143$ and $p/d = 1.43$ exceeds 1.0, whereas for $e/d = 0.057$ and $p/d = 1.43$, the friction factor was 0.13 only.

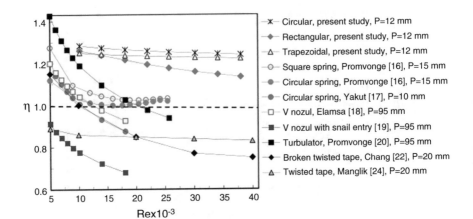

Fig. 6.35 Comparison of thermal performance in the grooved pipe with the results in the literature (Bilen et al. 2009)

Table 6.16 r_1 and r_2 for $Re = 10,633$

Tubes	Specifications	r_1	r_2
1	$p/d = 5.72$, $e/d = 0.075$	1.50	0.67
2	$p/d = 5.0$, $e/d = 0.05$	1.33	0.90
3	$p/d = 4.29$, $e/d = 0.075$	1.65	0.55
4	$p/d = 2.86$, $e/d = 0.057$	1.54	0.62
5	$p/d = 1.43$, $e/d = 0.143$	2.46	0.07
6	$p/d = 1.43$, $e/d = 0.057$	1.73	0.56
7	$p/d = 0.75$, $e/d = 0.015$	1.21	0.97
8	$p/d = 0.5$, $e/d = 0.0643$	2.10	0.32
9	$p/d = 0.304$, $e/d = 0.0286$	1.67	0.59

It signifies that e/d has direct influence on friction factor, but the overall influence of Reynolds number on the friction factor is very weak, and it varies slightly with Reynolds number. They plotted correlation results based on performance map of rib-roughened tubes and found a clear idea of that for a single p/d value, the r_1 increased with decreased value of p/d. The plot suggested that there was stronger effect of e/d on r_1 than that of p/d. It was observed that the effect of e/d and p/d on r_2 were just opposite to that on r_1. They concluded that entrance region has a slight effect on Nusselt number.

Aroonrat et al. (2013) examined experimentally the heat transfer and flow characteristics of water flowing in six tubes. The objective of the study was to identify the grooved pitch effect on heat transfer and flow characteristics. The different testing tubes were made up of stainless steel, and smooth tube (SMT), straight grooved tubes (SGT) and four helical grooved tube of different pitches named as GT 0.5 for 0.5-in. pitch, GT 8 for 8-in. pitch, GT 10 for 10-in. pitch and GT 12 for 12-in. pitch have been used. The details of test section were listed in Table 6.17.They calculated Nusselt number and friction factor at steady state

Table 6.17 Details of test section (Aroonrat et al. 2013)

Test tube	e (mm)	W (mm)	p (mm)	β (Degree)	d_o (mm)	d_i (mm)	N_s	A_c (mm²)	A_i (mm²)	L (mm)
SMT	–	–	–	–	9.5	7.1	–	39.6	44,611	2000
SGT	0.2	0.2	–	–	9.5	7.1	12	42	54,210	2000
GT0.5	0.2	0.2	12.7	60	9.5	7.1	10	41.6	61,497	2000
GT8	0.2	0.2	203	6.3	9.5	7.1	10	41.6	52,658	2000
GT10	0.2	0.2	254	5	9.5	7.1	10	41.6	52,641	2000
GT12	0.2	0.2	305	4.2	9.5	7.1	10	41.6	52,632	2000

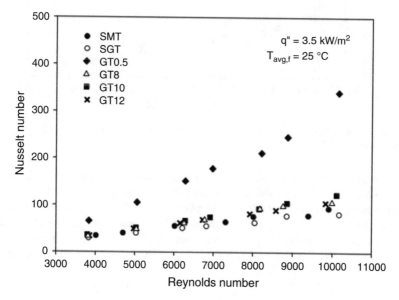

Fig. 6.36 Nusselt number as a function of Reynolds number for $q'' = 3.5$ kW/m² and $T_{avg, f} = 25$ °C (Aroonrat et al. 2013)

conditions. The Nusselt number versus Reynolds number data were plotted in Fig. 6.36 at average fluid temperature of 25 °C and at 3.5 kW/m² heat flux. The plotted graph clearly shows that that Nusselt number related to helical grooved tube was higher than the similar smooth tube. They observed the increase in Nusselt number as groove pitch decreases. They found that straight the Nusselt number of grooved tube was slightly less than that of smooth tube.

Thus, the straight grooved tube does not support enhancement, whereas a 260% enhancement is achieved from a 0.5-in. grooved tube and 25% for all other pitches in comparison to that for a smooth tube. They observed variation of friction factor with respect to Reynolds number and concluded that it decreased as the Reynolds number increased which was the expected result. They plotted the trend in Fig. 6.37 and calculated that 70% pressure loss was associated with 0.5-in. grooved tube.

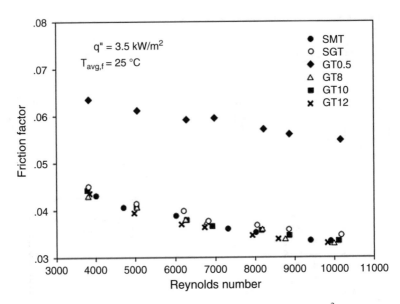

Fig. 6.37 Friction factor as a function of Reynolds number for $q'' = 3.5$ kW/m^2 and $T_{avg,f} = 25$ °C (Aroonrat et al. 2013)

However, other grooved tube friction factor performance was comparable to smooth tubes. They calculated thermal enhancement factor for all geometries and concluded that it was higher than that for all helical grooved tubes. The values were 1.4–2.2 for 0.5-in. pitch, 1.1–1.3 for 8-, 10- and 12-in. pitches as these pitches are comparable with each other. The straight groove tube thermal enhancement factor was 0.8–0.9 which is less than 1, which means it is inefficient in heat transfer enhancement.

References

Afroz HM, Miyara A (2007) Friction factor correlation and pressure loss of single-phase flow inside herringbone microfin tubes. Int J Refrig 30(7):1187–1194

Alam I, Ghoshdastidar PS (2002) A study of heat transfer effectiveness of circular tubes with internal longitudinal fins having tapered lateral profiles. Int J Heat Mass Transf 45:1371–1376

Al-Fahed SF, Ayub ZH, Al-Marafie AM, Soliman BM (1993) Heat transfer and pressure drop in a tube with internal microfins under turbulent water flow conditions. Exp Thermal Fluid Sci 7 (3):249–253

Al-Fahed S, Chamra LM, Chakroun W (1998) Pressure drop and heat transfer comparison for both microfin tube and twisted-tape inserts in laminar flow. Exp Thermal Fluid Sci 18(4):323–333

Arnold JA, Garimella S, Christensen RN (1993) Fluted tube heat exchanger design manual. GRI Report 5092-243-2357

Aroonrat K, Jumpholkul C, Leelaprachakul R, Dalkilic AS, Mahian O, Wongwises S (2013) Heat transfer and single-phase flow in internally grooved tubes. Int Commun Heat Mass Transf 42:62–68

Barba A, Bergles G, Gosman AD, Launder BE (1983) The prediction of convective heat transfer in viscous flow through spirally fluted tubes. ASME Paper 83-WA/HT-37

Bandarra Filho EP, Saiz Jabardo JM, Lopez Barbieri PE (2004) Convective boiling pressure drop of refrigerant R-134A in horizontal smooth and microfin tubes. Int J Refrig 27:895–903

Bar-Cohen A, Kraus AD (1990) Advances in thermal modeling of electronic components and systems. ASME, New York, p 2

Baughn JW, Kapila K, Perera CK, Yan X (1993) An experimental study of local heat transfer in a spirally fluted tube. In: Turbulent enhanced heat transfer, HTD, vol 239, pp 49–56

Bergles AE, Joshi SD (1983) Augmentation techniques for low Reynolds number in-tube flow. In: Kakaç S, Shah RK, Bergles AE (eds) Low Reynolds number flow heat exchangers. Hemisphere, Washington, DC, pp 694–720

Bergles AE, Manglik RM (2013) Current progress and new development in enhanced heat and mass transfer. J Enhanc Heat Transf 20(1):1–15

Bergman TL, Lavine AS, Incropera FP, Dewitt DP (2011) Introduction to heat transfer, 6th edn. Wiley, New York

Bhatia RS, Webb RL (2001) Numerical study of turbulent flow and heat transfer in microfin tubes— part 2, parametric study. J Enhanc Heat Transf 8:305–314

Bilen K, Cetin M, Gul H, Balta T (2009) The investigation of groove geometry effect on heat transfer for internally grooved tubes. Appl Therm Engg 29(4):753–761

Blumenkrantz A, Taborek J (1971) Heat transfer and pressure drop characteristics of Turbotec spirally deep grooved tubes in the laminar and transition regime. Report 2439-300-8, April 1971, Heat Transfer Research, Inc.

Bogart J, Thors P (1999) In-tube evaporation and condensation of R-22 and R-410A with plain and internally enhanced tubes. J Enhanc Heat Transf 6:37–50

Braga CVM, Saboya FEM (1986) Turbulent heat transfer and pressure drop in smooth and finned annular ducts. In: Heat transfer 1986, vol 6. Hemisphere Publishing Corporation, Washington, DC, pp 2831–2836

Brognaux LJ, Webb RL, Chamra LM, Chung BY (1997) Single-phase heat transfer in micro-fin tubes. Int J Heat Mass Transf 40:4345–4357

Carnavos TC (1979) Cooling air in turbulent flow with internally finned tubes. Heat Transf Eng 1 (2):41–46

Carnavos TC (1980) Heat transfer performance of internally finned tubes in turbulent flow. Heat Transf Eng 4(1):32–37

Cavallini A, Del Col D, Doretti L, Longo GA, Rossetto L (1997) Pressure drop during condensation and vaporization of refrigerants inside enhanced tubes. Heat Technol 15(1):3–10

Cavallini A, Del Col D, Mancin S, Rossetto L (2006) Thermal performance of R-410A condensing in a microfin tube. In: Proceedings of the international refrigeration conference at Purdue, R178

Celen A, Dalkilic AS, Wongwises S (2013) Experimental analysis of the single phase pressure drop characteristics of smooth and micro-fin tubes. Int Commun Heat Mass Transf 46:58–66

Chamra LM, Mago PJ (2007) Modeling of evaporation heat transfer of pure refrigerants and refrigerant mixtures in microfin tubes. Proc Inst Mech Eng, Part C J Mech Eng Sci 221:443–454

Chang SW, Jan YJ, Liou JS (2007) Turbulent heat transfer and pressure drop in tube fitted with serrated twisted tape. Int J Therm Sci 46(5):506–518

Chen J, Muller-Steinhagen H, Duffy GG (2001) Heat transfer enhancement in dimpled tubes. Appl Therm Eng 21:535–547

Choi JY, Kedzierski MA, Domanski PA (2001) Generalized pressure drop correlation for evaporation and condensation in smooth and microfin tubes. In: Proc of IIF-IIR Commission B1 Paderborn Germany, vol B4, pp 9–16

Choi JM, Kim Y, Lee M (2010) Air side heat transfer coefficients of discreteplate finned-tube heat exchangers with large fin pitch. Appl Therm Eng 30(s2–3):174–180

Choudhury D, Patankar SV (1985) Analysis of developing laminar flow and heat transfer in tubes with radial internal fins. In: Shenkman SM, O'Brien JE, Habib IS, Kohler JA (eds) Advances in enhanced heat transfer, HTD, vol 43, pp 57–64

Collier JG, Thome JR (1994) Convective boiling and condensation, 3rd edn. Oxford University Press, Oxford

Cope WG (1945) The friction and heat transmission coefficients of rough pipes. Proc Inst Mech Eng 145:99–105

Copetti JB, Macagnan MH, de Souza D, Oliveski RDC (2004) Experiments with micro-fin tube in single phase. Int J Refrig 27(8):876–883

Dagtekin I, Oztop HF, Sahin AZ (2005) An analysis of entropy generation through a circular duct with different shaped longitudinal fins for laminar flow. Int J Heat Mass Transf 48:171–181

Dipprey DF, Sabersky RH (1963) Heat and momentum transfer in smooth and rough tubes at various Prandtl number. Int J Heat Mass Transf 6:329–353

Duan L, Ling X, Peng H (2018) Flow and heat transfer characteristics of a double-tube structure internal finned tube with blossom shape internal fins. Appl Therm Eng 128:1102–1115

Eckert ERG, Irvine TF (1960) Pressure drop and heat transfer in a duct with triangular cross section. J Heat Transf 82(2):125–136

El-Sayed SA, Abdel-Hamid ME, Sadoun MM (1997) Experimental study of turbulent flow inside a circular tube with longitudinal interrupted fins in the streamwise direction. Exp Therm Fluid Sci 15:1–15

Fabbri G (1998) Heat transfer optimization in internally finned tubes under laminar flow conditions. Int J Heat Mass Transf 41(10):1243–1253

Fabbri G (1999) Optimum profiles for asymmetrical longitudinal fins in cylindrical ducts. Int J Heat Mass Transf 42:511–523

Fabbri G (2004) Effect of viscous dissipation on the optimization of the heat transfer in internally finned tubes. Int J Heat Mass Transf 47:3003–3015

Fabbri G (2005) Optimum cross-section design of internally finned tubes cooled by a viscous fluid. Control Eng Pract 13:929–938

Fujie K, Itoh N, Innami T, Kimura H, Nakayama N, Yanugidi T (1977) Heat transfer pipe. U. S. Patent 4,044,797, assigned to Hitachi Ltd

García A, Solano JP, Vicente PG, Viedma A (2012) The influence of artificial roughness shape on heat transfer enhancement: corrugated tubes, dimpled tubes and wire coils. Appl Therm Eng 35:196–201

Garimella S, Christensen RN (1995a) Heat transfer and pressure drop characteristics of spirally fluted annuli: part I—hydrodynamics. J Heat Transf 117:54–60

Garimella S, Christensen RN (1995b) Heat transfer and pressure drop characteristics of spirally fluted annuli: park II—heat transfer. J Heat Transf 117:61–68

Gee DL, Webb RL (1980) Forced convection heat transfer in helically rib-roughened tubes. Int J Heat Mass Transf 23(8):1127–1136

Ghiaasiaan SM (2008) Two-phase flow boiling and condensation. Cambridge University Press, Cambridge

Goto M, Inoue N, Ishiwatari N (2001) Condensation and evaporation heat transfer of R-410A inside internally grooved horizontal tubes. Int J Refrig 24(7):628–638

Goto M, Inoue N, Yonemoto R (2003) Condensation heat transfer of R410A inside internally grooved horizontal tubes. Int J Refrig 26(4):410–416

Gowen RA, Smith JW (1968) Turbulent heat transfer from smooth and rough surfaces. Int J Heat Mass Transf 11:1657–1673

Hamilton LJ, Kedzierski MA, Kaul MP (2008) Horizontal convective boiling of pure and mixed refrigerants within a micro-fin tube. J Enhanc Heat Transf 15(3):211–226

Han DH, Lee KJ (2005) Single-phase heat transfer and flow characteristics of micro-fin tubes. Appl Therm Engg 25(11–12):1657–1669

Hatami M, Jafaryar M, Ganji DD, Gorji-Bandpy M (2014) Optimization of finned-tube heat exchangers for diesel exhaust waste heat recovery using CFD and CCD techniques. Int Commun Heat Mass 57:254–263

Hatami M, Ganji DD, Gorji-Bandpy M (2015) Experimental and numerical analysis of the optimized finned-tube heat exchanger for OM314 diesel exhaust exergy recovery. Energy Convers Manag 97:26–41

Hilding WE, Coogan CH Jr (1964) Heat transfer and pressure drop in internally finned tubes. In: ASME symposium on air cooled heat exchangers. ASME, New York, pp 57–84

Hu MH, Chang YP (1973) Optimization of finned tubes for heat transfer in laminar flow. J Heat Transf 95(3):332–338

Huang D, Zhao RJ, Liu Y (2014) Effect of fin types of outdoor fan-supplied finned-tube heat exchanger on periodic frosting and defrosting performance of a residential air-source heat pump. Appl Therm Eng 69(1–2):251–260

Huq M, Huq AAU, Rahman MM (1998) Experimental measurements of heat transfer in an internally finned tube. Int Commun Heat Mass Transf 25(5):619–630

Iqbal Z, Syed KS, Ishaq M (2013) Optimal fin shape in finned double pipe with fully developed laminar flow. Appl Therm Eng 51:1202–1223

Islam MA, Mozumder AK (2009) Forced convection heat transfer performance of an internally finned tube. J Mech Eng 40:54–62

Ivanović M, Selimović R, Bajramović R (1990) Mathematical modeling of heat transfer in internally finned tubes. In: Hanjalić H (ed) Mathematical modeling and computer simulation of processes in energy systems. Hemisphere Publishing Corp, Washington, DC, pp 147–153

Jensen MK, Vlakancic A (1999) Technical note experimental investigation of turbulent heat transfer and fluid flow in internally finned tubes. Int J Heat Mass Transf 42(7):1343–1351

Kelkar KM, Patankar SV (1990) Numerical prediction of fluid flow and heat transfer in a circular tube with longitudinal fins interrupted in the steamwise direction. J Heat Transf 112:342–348

Kido O, Taniguchi M, Taira T, Uehara H (1995) Evaporation heat transfer of HCFC22 inside an internally grooved horizontal tube. Proc ASME/JSME Therm Eng Conf 2:323–330

Kim NH (2015a) Single-phase pressure drop and heat transfer measurements of turbulent flow inside helically dimpled tubes. J Enhanc Heat Transf 22(4):345–363

Kim NH (2015b) Effect of aspect ratio on evaporation heat transfer and pressure drop of R-410A in flattened microfin tubes. J Enhanc Heat Transf 22(3):177–197

Kim DK (2016) Thermal optimization of internally finned tube with variable fin thickness. Appl Therm Eng 102:1250–1261

Kim DK, Kim SJ (2007) Closed form correlations for thermal optimization of microchannels. Int J Heat Mass Transf 50(25):5318–5322

Kim NH, Kim SH (2010) Dry and wet air-side performance of a louver-finned heat exchanger having flat tubes. J Mech Sci Technol 24:1553–1561

Kim NH, Webb RL (1989) Experimental study of particulate fouling in enhanced water chiller condenser tubes. ASHRAE Trans 76(2):507–515

Kim NH, Webb RL (1993) Analytic prediction of the friction and heat transfer for turbulent flow in axial internal fin tubes. J Heat Transf 115(3):553–559

Kim SJ, Yoo JW, Jang SP (2002) Thermal optimization of a circular-sectored finned tube using a porous medium approach. J Heat Transf 124(6):1026–1033

Kim DK, Jung J, Kim SJ (2010) Thermal optimization of plate-fin heat sinks with variable fin thickness. Int J Heat Mass Transf 53(25):5988–5995

Kim NH, Lee EJ, Byun HW (2013) Evaporation heat transfer and pressure drop of R-410A in flattened smooth tubes having different aspect ratios. Int J Refrig 36:363–374

Kiml R, Magda A, Mochizuki S, Murata A (2004) Rib-induced secondary flow effects on local circumferential heat transfer distribution inside a circular rib-roughened tube. Int J Heat Mass Transf 47(6–7):1403–1412

Koyama S, Yu J, Momoki S, Fujii T, Honda H (1995) Forced convective flow boiling heat transfer of pure refrigerants inside a horizontal microfin tube. In: Proceedings of the engineering foundation conference on convective flow boiling. ASME Banff Canada

Kumar R (1997) Three-dimensional natural convective flow in a vertical annulus with longitudinal fins. Int J Heat Mass Transf 40:3323–3334

Kumbhar DG, Sane NK (2015) Exploring heat transfer and friction factor performance of a dimpled tube equipped with regularly spaced twisted tape inserts. Procedia Eng 127:1142–1149

Kuo CS, Wang CC (1996) In-tube evaporation of HCFC-22 in a 9.52 mm micro-fin/smooth tube. Int J Heat Mass Transf 39(12):2559–2569

Kuwahara H, Takahashi K, Yanagida T, Nakayama W, Sugimoto S, Oizumi K (1989) Method of producing a heat transfer tube for single-phase flow. US Patent 4,794,775 issued to Hitachi Cable Ltd

Lemouedda A, Schmid A, Franz E et al (2011) Numerical investigations for the optimization of serrated finned-tube heat exchangers. Appl Therm Eng 31(8–9):1393–1401

Li XW, Meng JA, Li ZX (2007) Experimental study of single-phase pressure drop and heat transfer in a micro-fin tube. Exp Thermal Fluid Sci 32(2):641–648

Li GQ, Wu Z, Li W, Wang ZK, Wang X, Li HX, Yao SC (2012) Experimental investigation of condensation in micro-fin tubes of different geometries. Exp Thermal Fluid Sci 37:19–28

Liao Q, Xin XD (1995) Experimental investigation on forced convective heat transfer and pressure drop of ethylene glycol in tubes with three-dimensional internally extended surface. Exp Therm Fluid Sci 11:343–347

Liao Q, Xin XD (2000) Augmentation of convective heat transfer inside tubes with three dimensional internal extended surfaces and twisted-tape inserts. Chem Eng J 78:95–105

Liao Q, Zhu X, Xin MD (2000) Augmentation of turbulent convective heat transfer in tubes with three-dimensional internal extended surfaces. J Enhanc Heat Transf 7(3):139–151

Lin ZM, Wang LB, Zhang YH (2014) Numerical study on heat transfer enhancement of circular tube bank fin heat exchanger with interrupted annular groove fin. Appl Therm Eng 73:1465–1476

Liu XY, Jensen MK (1999) Numerical investigation of turbulent flow and heat transfer in internally finned tubes. J Enhanc Heat Transf 6:105–119

Liu X, Jensen MK (2001) Geometry effects on turbulent flow and heat transfer in internally finned tubes. J Heat Transf 123(6):1035–1044

Liu L, Ling X, Peng H (2013a) Complex turbulent flow and heat transfer characteristics of tubes with internal longitudinal plate-rectangle fins in EGR cooler. Appl Therm Eng 54:145–152

Liu L, Fan YZ, Ling X, Peng H (2013b) Flow and heat transfer characteristics of finned tube with internal and external fins in air cooler for waste heat recovery of gas-fired boiler system. Chem Eng Process 74:142–152

Liu L, Ling X, Peng H (2015) Study on turbulent flow and heat transfer performance of tubes with internal fins in EGR cooler. Heat Mass Transf 1:1017–1027

Luo YM, Shao SQ, Xu HB, Tian CQ, Yang HX (2014) Experimental and theoretical research of a fin-tube type internally-cooled liquid desiccant dehumidifier. Appl Energy 133:127–134

Ma Y, Yuan Y, Liu Y (2012) Experimental investigation of heat transfer and pressure drop in serrated finned tube banks with staggered layouts. Appl Therm Eng 37(5):314–323

Mahmood GI, Ligrani PM (2002) Heat transfer in a dimpled channel: combined influences of aspect ratio, temperature ratio, Reynolds number and flow structure. Int J Heat Mass Transf 45 (10):2011–2020

Manglik RM, Bergles AE (1993) Heat transfer and pressure drop correlations for twisted-tape inserts in isothermal tubes: part 1—laminar flows. J Heat Transf 115(4):881–889

Marner WJ, Bergles AE (1978) Augmentation of tubeside laminar flow heat transfer by means of twisted-tape inserts, static-mixer inserts, and internally finned tubes. In: Heat transfer 1978, vol 2. Hemisphere Publishing Corporation, Washington, DC, pp 583–588

Marner WJ, Bergles AE (1985) Augmentation of highly viscous laminar tubeside heat transfer by means of a twisted tape insert and an internally finned tube. In: Shenkman SM, O'Brien JE, Habib IS, Kohler JA (eds) Advances in enhanced heat transfer, HTD, vol 43, pp 19–28

Marner WJ, Bergles AE (1989) Augmentation of highly viscous laminar heat transfer inside tubes with constant wall temperature. Exp Therm Fluid Sci 2:252–267

Martinez E, Vicente W, Salinas-Vazquez M, Carvajal I, Alvarez M (2015) Numerical simulation of turbulent air flow on a single isolated finned tube module with periodic boundary conditions. Int J Therm Sci 92:58–71

Marto PJ, Reilly DJ, Fenner JH (1979) An experimental comparison of enhanced heat transfer condenser tubing. In: Advances in enhanced heat transfer. ASME, New York, pp 1–9

Moreno Quiben J, Cheng L, da Silva Lima RJ, Thome JR (2009a) Flow boiling in horizontal flatten tubes: part I—two-phase frictional pressure drop results and model. Int J Heat Mass Transf 52:3634–3644

Moreno Quiben J, Cheng L, da Silva Lima RJ, Thome JR (2009b) Flow boiling in horizontal flattened tubes: part II—flow boiling heat transfer results and model. Int J Heat Mass Transf 52:3645–3653

Mukkamala Y, Sundaresan R (2009) Single-phase flow pressure drop and heat transfer measurements in a horizontal microfin tube in the transition regime. J Enhanc Heat Transf 16 (2):141–159

Nandakumar K, Masliyah HH (1975) Fully developed viscous flow in internally finned tubes. Chem Eng J 10:113–120

Nasr M, Akhavan-Behabadi MA, Marashi SE (2010) Performance evaluation of flattened tube in boiling heat transfer enhancement and its effect on pressure drop. Int Commun Heat Mass Transf 37:430–436

Newell TA, Shah RK (2001) An assessment of refrigerant heat transfer, pressure drop and void fraction effects in microfin tubes. Int J HVAC&R Res 7(2):125–153

Nikuradse J (1922) Law of flows in rough pipes, Forsh Arb Ing—Wesen No 361 Translated NACATM 1292 (1950)

Nishida S, Murata A, Saito H, Iwamoto K (2012) Compensation of three-dimensional heat conduction inside wall in heat transfer measurement of dimpled surface by using transient technique. J Enhanc Heat Transf 19(4):331–341

Nivesrangsan P, Pethkool S, Nanan K, Pimsarn M, Eiamsa-ard S (2010) Thermal performance assessment of turbulent flow through dimpled tubes. In: Proc. 14th international heat transfer conference IHTC14-22503 Washington, DC

Obot NT, Esen EB, Snell KH, Rabas TJ (1991) Pressure drop and heat transfer for spirally fluted tubes including validation of the role of transition. In: Rabas TJ, Chenoweth JM (eds) Fouling and enhancement interactions, ASME Symp. HTD, vol 164, pp 85–92

Olson DA (1992) Heat transfer in thin, compact heat exchangers with circular, rectangular, or pin-fin flow passages. J Heat Transf 114:373–382

Panchal CB, France DM (1986) Performance tests of the spirally fluted tube heat exchanger for industrial cogeneration applications. Argonne National Laboratory Report ANL/CNSV-59

Park J, Ligrani PM (2005) Numerical predictions of heat transfer and fluid flow characteristics for seven different dimpled surfaces in a channel. Numer Heat Transf Part A Appl 47(3):209–232

Patankar SV, Chai JC (1991) Laminar natural convection in internally finned horizontal annuli. ASME Paper No. 91-HT-12

Patankar SV, Ivanovic M, Sparrow EM (1979) Analysis of turbulent flow and heat transfer in internally finned tubes and annuli. ASME J Heat Transf 101:29–37

Peng H, Ling X (2011) Analysis of heat transfer and flow characteristics over serrated fins with different flow directions. Energy Convers Manag 52:826–835

Peng H, Ling X, Li J (2014) Performance investigation of an innovative offset strip fin arrays in compact heat exchangers. Energy Convers Manag 80:287–297

Peng H, Liu L, Ling X, Li Y (2016) Thermo-hydraulic performances of internally finned tube with a new type wave fin arrays. Appl Therm Eng 98:1174–1188

Perera KK, Baughn JW (1994) The effect of pitch angle and Reynolds number on local heat transfer in spirally fluted tubes. In: Haas LA, Downing RS (eds) Optimal design of thermal systems and components, HTD, vol 279, pp 99–112

Prakash C, Liu Y-D (1985) Analysis of laminar flow and heat transfer in the entrance region of an internally finned circular duct. J Heat Transf 107:84–91

Prakash C, Patankar SV (1981) Combined free and forced convection in internally finned tubes with radial fins. J Heat Transf 103:566–572

Promvonge P, Eiamsa-Ard S (2007) Heat transfer augmentation in a circular tube using V-nozzle turbulator inserts and snail entry. Exp Therm Fluid Sci 32(1):332–340

Promvonge P (2015) Thermal performance in square-duct heat exchanger with quadruple V-finned twisted tapes. Appl Therm Eng 91:298–307

Rabas TJ, Mitchell H (2000) Internally enhanced carbon steel tubes for ammonia chillers. Heat Transf Eng 21(5):3–16

Raj R, Lakshman NS, Mukkamala Y (2015) Single phase flow heat transfer and pressure drop measurements in doubly enhanced tubes. Int J Therm Sci 88:215–227

Ravigururajan TS, Bergles AE (1995) Prandtl number influence on heat transfer enhancement in turbulent flow of water at low temperatures. J Heat Transf 117(2):276–282

Richards DE, Grant MM, Christensen RN (1987) Turbulent flow and heat transfer inside doubly-fluted tubes. ASHRAE Trans 93(Part 2):2011–2026

Rout SK, Thatoi DN, Acharya AK, Mishra DP (2012) CFD supported performance estimation of an internally finned tube heat exchanger under mixed convection flow. Procedia Eng 38:585–597

Rowley GJ, Patankar SV (1984) Analysis of laminar flow and heat transfer in tubes with internal circumferential fins. Int J Heat Mass Transf 27(4):553–560

Rustum IM, Soliman HM (1988a) Experimental investigation of laminar mixed convection in tubes with longitudinal internal fins. J Heat Transf 110:366–372

Rustum IM, Soliman HM (1988b) Numerical analysis of laminar forced convection in the entrance region of tubes with longitudinal internal fins. J Heat Transf 110:310–313

Rustum IM, Soliman HM (1990) Numerical analysis of laminar mixed convection in horizontal internally finned tubes. Int J Heat Mass Transf 33(7):1485–1496

Saad AE, Sayed AE, Mohamed EA, Mohamed MS (1997) Experimental study of turbulent flow inside a circular tube with longitudinal interrupted fins in the streamwise direction. Exp Therm Fluid Sci 15(1):1–15

Said NMA, Trupp AC (1984) Predictions of turbulent flow and heat transfer in internally finned tubes. Chem Eng Commun 31:65–99

San JY, Huang WC (2006) Heat transfer enhancement of transverse ribs in circular tubes with consideration of entrance effect. Int J Heat Mass Transf 49(17–18):2965–2971

Sarkhi A, Nada E (2005) Characteristics of forced convection heat transfer in vertical internally finned tube. Int Commun Heat Mass 32:557–564

Shih TH, Liou WW, Shabbrir A, Yang ZG, Zhu J (1995) A new k–e eddy viscosity model for high Reynolds number turbulent flows. Comput Fluids 24(3):227–238

Shome B (1998) Mixed convection laminar flow and heat transfer of liquids in horizontal internally finned tubes. Numer Heat Transf Part A 33(1):65–84

Shome B, Jensen MK (1996a) Experimental investigation of laminar flow and heat transfer in internally finned tubes. J Enhanc Heat Transf 4:53–70

Shome B, Jensen MK (1996b) Numerical investigation of laminar flow and heat transfer in internally finned tubes. J Enhanc Heat Transf 4:35–52

Siddique M, Alhazmy M (2008) Experimental study of turbulent single-phase flow and heat transfer inside a micro-finned tube. Int J Refrig 31(2):234–241

Soliman HM (1979) The effect of fin material on laminar heat transfer characteristics of internally finned tubes. In: Chenoweth JM, Kaellis J, Michel JW, Shenkman S (eds) Advances in enhanced heat transfer. ASME, New York, pp 95–102

Soliman HM, Feingold A (1977) Analysis of fully developed laminar flow in longitudinally internally finned tubes. Chem Eng J 14:119–128

Soliman HM, Chau TS, Trupp AC (1980) Analysis of laminar heat transfer in internally finned tubes with uniform outside wall temperature. J Heat Transf 102:598–604

Song WM, Meng JA, Li ZX (2010) Numerical study of air-side performance of a finned flat tube heat exchanger with crossed discrete double inclined ribs. Appl Therm Eng 30(13):1797–1804

Sparrow EM, Lovell B (1980) Heat transfer characteristics of an obliquely impinging circular jet. J Heat Transf 102(2):202–209

Srinivasan V, Christensen RN (1992) Experimental investigation of heat transfer and pressure drop characteristics of flow through spirally fluted tubes. Exp Therm Fluid Sci 5:820–827

Srinivasan V, Vafai K, Christensen RN (1994) Experimental investigation, modeling and prediction of friction factors and friction increase ratio for flow through spirally fluted tubes. J Enhanc Heat Transf 1(4):337–350

Suresh S, Chandrasekar M, Chandrasekar S (2001) Experimental studies on heat transfer and friction factor characteristics of CuO/water nanofluid under turbulent flow in a helically dimpled tube. Exp Thermal Fluid Sci 35:542–549

Syed KS, Ishaq M, Iqbal Z, Hassan A (2015) Numerical study of an innovative design of a finned double-pipe heat exchanger with variable fin-tip thickness. Energy Convers Manag 98:69–80

Takahashi K, Nakayama W, Kuwahara H (1988) Enhancement of forced convective transfer in tubes having three-dimensional spiral ribs. Heat Transf Jpn Res 17(4):12–28

Thianpong C, Eiamsa-ard P, Wongcharee K, Eiamsa-ard S (2009) Compound heat transfer enhancement of a dimpled tube with a twisted tape swirl generator. Int Commun Heat Mass Transf 36:698–704

Thome JR, Kattan N, Favrat D (1997) Evaporation in micro-fin tubes: a generalized prediction model. In: Proc. convective flow and pool boiling conference, Kloster Irsee (Paper VII-4)

Trupp AC, Haine H (1989) Experimental investigation of turbulent mixed convection in horizontal tubes with longitudinal internal fins. In: Shah RK (ed) Heat transfer in convective flows, ASME HTD, vol 107, pp 17–25

Trupp AC, Lau ACY, Said NNA, Soliman HM (1981) Turbulent flow characteristics in an internally finned tube. In: Webb RL, Carnavos TC, Park EL Jr, Hostetler KM (eds) Advances in enhanced heat transfer 1981, ASME Symp. HTD, vol 18. ASME, New York, pp 11–20

Wang C-C, Chen P-Y, Jang J-Y (1996) Heat transfer and friction characteristics of convex-louver fin-and-tube heat exchangers. Exp Heat Transf 9:61–78

Wang QW, Lin M, Zeng M, Tian L (2008a) Computational analysis of heat transfer and pressure drop performance for internally finned tubes with three different longitudinal wavy fins. Heat Mass Transf 45:147–156

Wang QW, Lin M, Zeng M, Tian L (2008b) Investigation of turbulent flow and heat transfer in periodic wavy channel of internally finned tube with blocked core tube. J Heat Transf 130(6). Article No.: 061801

Wang QW, Lin M, Zeng M (2009) Effect of lateral fin profiles on turbulent flow and heat transfer performance of internally finned tubes. Appl Therm Eng 29:3006–3013

Wang Y, He Y-L, Lei Y-G, Zhang J (2010) Heat transfer and hydrodynamics of a novel dimpled tube. Exp Therm Fluid Sci 34:1273–1281

Wang YG, Zhao QX, Zhou QL, Kang ZJ, Tao WQ (2013) Experimental and numerical studies on actual flue gas condensation heat transfer in a left-right symmetric internally finned tube. Int J Heat Mass Transf 64:10–20

Wang QW, Zeng M, Ma T, Du XP, Yang JF (2014) Recent development and application of several high-efficiency surface heat exchangers for energy conversion and utilization. Appl Energy 135:748–777

Wang WJ, Bao Y, Wang YQ (2015) Numerical investigation of a finned-tube heat exchanger with novel longitudinal vortex generators. Appl Therm Eng 86:27–34

Wang YH, Zhang JL, Ma ZX (2017) Experimental determination of single-phase pressure drop and heat transfer in a horizontal internal helically-finned tube. Int J Heat Mass Transf 104:240–246

Watkinson AP, Miletti PL, Tarassoff p (1973) Turbulent heat transfer and pressure drop in internally finned tubes. AIChE Symp Ser 69(131):94–103

Watkinson AP, Miletti PL, Kubanek GR (1975a) Heat transfer and pressure drop of internally finned tubes in laminar oil flow. ASME Paper 75-HT-41

Watkinson AP, Miletti PL, Kubanek GR (1975b) Heat transfer and pressure drop of internally finned tubes in turbulent air flow. ASHRAE Trans 81(Part 1):330–349

Webb RL (1981) The use of enhanced heat transfer surface geometries in condensers. In: Marto PJ, Nunn RH (eds) Power condenser heat transfer technology: computer modeling, design, fouling. Hemisphere Pub. Corp., Washington, DC, pp 287–324

Webb RL, Iyengar A (2001) Oval finned tube condenser and design pressure limits. J Enhanc Heat Transf 8:147–158

Webb RL, Kim NH (2005) Principles of enhanced heat transfer, 2nd edn. Taylor & Francis, London

Webb RL, Scott MJ (1980) A parametric analysis of the performance of internally finned tubes for heat exchanger application. J Heat Transf 102(1):38–43

Webb RL, Eckert ERG, Goldstein R (1971) Heat transfer and friction in tubes with repeated-rib roughness. Int J Heat Mass Transf 14(4):601–617

Webb RL, Narayanamurthy R, Thors P (2000) Heat transfer and friction characteristics of internal helical-rib roughness. J Heat Transf 122(1):134–142

Wilson MJ, Newell TA, Chato JC, Infante Ferreira CA (2003) Refrigerant charge, pressure drop and condensation heat transfer in flattened tubes. Int J Refrig 26:442–451

Wolfstein M (1988) The velocity and temperature distribution of one dimensional flow with turbulence augmentation and pressure gradient. Int J Heat Mass Transf 12:301–318

Wu Z, Wu Y, Sunden B, Li W (2013) Convective vaporization in micro-fin tubes of different geometries. Exp Thermal Fluid Sci 44:398–408

Yakut K, Sahin B, Canbazoglu S (2004) Performance and flow-induced vibration characteristics for conical-ring turbulators. Appl Energy 79(1):65–76

Yampolsky JS (1983) Spirally fluted tubing for enhanced heat transfer. In: Taborek J, Hewitt GF, Afgan N (eds) Heat exchangers-theory and practice. Hemisphere Publishing Corp, Washington, DC, pp 945–952

Yu B, Nie JH, Wang QW, Tao WQ (1999) Experimental study on the pressure drop and heat transfer characteristics of tubes with internal wave-like longitudinal fins. Heat Mass Transf 35:65–73

Yu B, Tao WQ (2004) Pressure drop and heat transfer characteristics of turbulent flow in annular tubes with internal wave-like longitudinal fins. Heat Mass Transf 40:643–651

Yun R, Kim Y, Seo K, Kim HY (2002) A generalized correlation for evaporation heat transfer of refrigerants in micro-fin tubes. Int J Heat Mass Transf 45:2003–2010

Zdaniuk GJ, Luck R, Chamra LM (2008) Linear correlation of heat transfer and friction in helically-finned tubes using five simple groups of parameters. Int J Heat Mass Transf 51 (13–14):3548–3555

Zeitoun O, Hegazy AS (2004) Heat transfer for laminar flow in internally finned pipes with different fin heights and uniform wall temperature. Heat Mass Transf 40:253–259

Zhang HY, Ebadian MA (1992a) Heat transfer in the entrance region of semicircular ducts with internal fins. J Thermophys Heat Transf 6:296–301

Zhang HY, Ebadian MA (1992b) The influence of internal fins on mixed convection inside a semicircular duct. In: Pate MB, Jensen MK (eds) Enhanced heat transfer. ASME Symp. HTD, vol 202. ASME, New York, pp 17–24

Zhang HG, Wang EH, Fan BY (2013) Heat transfer analysis of a finned-tube evaporator for engine exhaust heat recovery. Energy Convers Manag 65:438–447

Chapter 7
Advanced Internal Fin Geometries and Finned Annuli

Microfins may be formed in a copper tube at high speed by drawing the tube over a grooved slug. This is applied for convective vaporization and condensation of refrigerants. Khanpara et al. (1987) measured the enhancement of subcooled R22 and R113 liquid flow in the microfin tube. Koyama et al. (1996) measured the heat transfer coefficient for heating of water and subcooled liquid refrigerants R22, R123 and R134a in a microfin tube (Fig. 7.1).

Li et al. (2008) reported the performance of microfins for single-phase heat transfer enhancement in turbulent flow regime. They carried out a numerical investigation for tube having helical microfins and compared that having straight microfins. They observed that the critical Reynolds number for flow microfin tubes is dependent on the ratio of viscous sublayer thickness to the fin roughness height. The roughness elements do not cause considerable turbulence and fluid flow mixing for heights less than viscous boundary layer thickness.

The heat transfer enhancement is achieved only when the height of the roughness is greater than the viscous boundary layer thickness, by disturbing the boundary layer formation and augmenting the flow mixing. The increase in both heat transfer and friction factor was observed for increasing fin height, which establishes the need for optimum fin height which maximizes the overall efficiency of the fin. For high efficiency of the fin, additional turbulence at the wall is required with minimum form drag. Also, for greater Reynolds number, the shear stress friction factor is low for the helical microfin tube. Thus, the friction factor for helical microfin tubes decreases with increase in Reynolds number.

Jensen and Vlakancic (1999) reported that microfin tubes have $e/d_i \leq 0.03$ and $n \geq 30$, where n is the number of fins. Webb and Kim (2005) presented the use of microfin tubes for two-phase heat transfer enhancement. Shedd and Newell (2003) and Shedd et al. (2003) have also worked on two-phase flow in microfin tubes. Li et al. (2007) presented a thorough review on microfin tubes. They observed that a critical Reynolds number exists for heat transfer enhancement in microfin tubes. Also, they reported that heat transfer performance in microfin is entirely different

© The Author(s), under exclusive license to Springer Nature Switzerland AG 2020
S. K. Saha et al., *Heat Transfer Enhancement in Externally Finned Tubes and Internally Finned Tubes and Annuli*, SpringerBriefs in Applied Sciences and Technology, https://doi.org/10.1007/978-3-030-20748-9_7

Fig. 7.1 Single-phase
microfin heat transfer data
predicted by single-phase
heat transfer correlations
(Koyama et al. 1996)

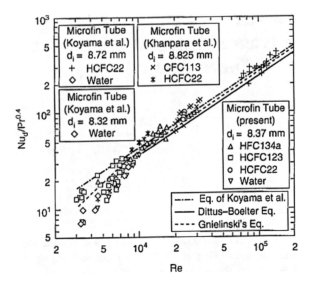

Table 7.1 Geometric
parameters of the test tubes
(Han and Lee 2005)

Tube no.	1	2	3	4
D_0 (mm)	9.52	7	6.2	5.1
t (mm)	0.3	0.27	0.55	0.55
e (mm)	0.12	0.15	0.13	0.13
p (mm)	1	1.04	1.47	1.32
β (rad)	0.44	0.31	0.18	0.16
n	60	60	60	60

from that of internal rough tubes. Liu and Jensen (2001) carried out numerical work
on microfin tubes using k-ε turbulence model.

Han and Lee (2005) investigated the heat transfer coefficients and flow characteristics of microfin tube and developed correlations based on experimental data.
They used horizontal double-pipe heat exchanger with water as the working fluid.
They measured the heat transfer and pressure drop in the range 15–45 °C with
Reynolds number range 3000–40,000 and Prandtl number Pr from 4 to 7. The
specifications are elaborated in Table 7.1. It is well known that spiral angle of fin
(β), fin height (e), fin pitch (p) and number of fins (n) are parameters that significantly affect the heat transfer augmentation. Thus, Han and Lee (2005) developed
correlation for friction factor taking p/e factor into consideration as

$$f = 0.19 Re^{-0.024}(p/e)^{0.54} \tag{7.1}$$

It shows that pressure drop increases with increase in fin height (e/d) and
decreases with pitch (p/d). The microfin tube is shown in Fig. 7.2. They also
introduced non-dimensional temperature t^{+} and developed the correlation as

Fig. 7.2 Microfin tube geometry (Han and Lee 2005)

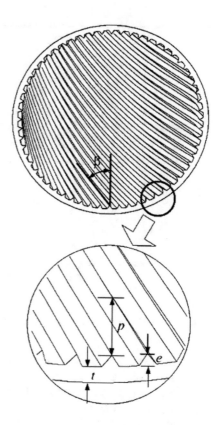

$$\Delta t_w^+ Pr^{-0.55} = 12.47[0.1705(a - Re_\varepsilon)] - 0.1147(a - Re_\varepsilon) + b \qquad (7.2)$$

$$a = \left[592.91 - 204.8 \ln\left(\frac{p}{e}\right)\right]\left(\frac{e}{D_i}\right)^{0.74} \qquad (7.3)$$

$$b = 0.5849 + 5.8174 \ln a \qquad (7.4)$$

To conclude the effectiveness of enhancement techniques, they defined efficiency index

$$\eta = \frac{\frac{hA}{h_s A_s}}{\frac{f}{f_s}} \qquad (7.5)$$

and predicted the values with Reynolds number presented in Fig. 7.3. It can be observed from this plot that higher relative roughness and smaller spiral angle, as in tube 3 and tube 4, enhance the heat transfer coefficients significantly in comparison

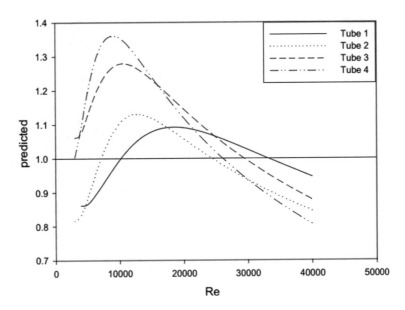

Fig. 7.3 Calculated efficiency index vs. *Re* (Han and Lee 2005)

to smaller relative roughness and larger spiral angle of tube (tube 1). Their predicted mean deviation was 6.4%.

Tam et al. (2012) investigated the characteristics of heat transfer, friction factor and optimal fin geometries for the internally microfin tubes in the laminar, transition and turbulent regions. They carried out an experimental investigation on three microfin tubes with different inlet configurations such as squared-edge and re-entrant and compared the experimental results with the data of plain tubes. It was observed that efficiency index became more than 1 when the Reynolds number was larger than 5000 for all the microfin tubes with both inlet configurations. Figure 7.4 shows the schematic diagram of plain and microfin tubes. They solved the genetic algorithms and the algorithms of changes with the help of existing turbulent correlation, so that fin geometries could be optimized. They obtained the experimental results for friction factor and heat transfer for horizontal plain and microfin tubes simultaneously under isothermal and uniform wall heat flux boundary conditions.

Table 7.2 shows the geometrical parameters of test tubes. Table 7.3 shows the values of the friction factor and heat transfer for different Reynolds number at the start and end of the transition of all microfin and plain tubes under the isothermal and heating conditions. Figure 7.5 shows the variation of friction factor for all the tubes, in the range from laminar to turbulent regions under isothermal boundary conditions. It was observed that beginning of transition depends on both inlet and spiral angle. It was found that transition started earlier and also ended early at larger spiral angle of the microfin. Figure 7.6 shows the variation of Colburn *j* factor ($St\, Pr^{0.67}$) for the plain and microfin tubes. Khanpara et al. (1986), Brognaux et al. (1997), Jensen and

Fig. 7.4 (a) Sectional view
of the microfin tube;
(b) plain and microfin
tubes (Tam et al. 2012)

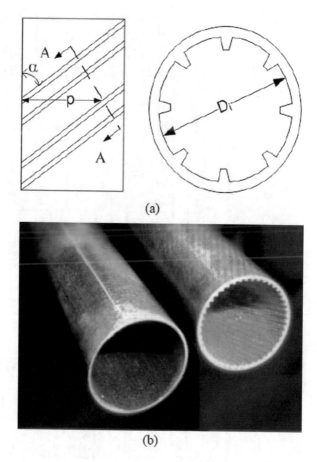

(a)

(b)

Table 7.2 Specifications of the test tubes (Tam et al. 2012)

Tube type	Outer diameter, D_0 (mm)	Inner diameter, D_i (mm)	Spiral angle, α	Fin height, e (mm)	Number of starts, N_s
Plain tube	15.9	14.9	–	–	–
Microfin tube #1	15.9	14.9	18°	0.5	25
Microfin tube #2	15.9	14.9	25°	0.5	25
Microfin tube #3	15.9	14.9	35°	0.5	25

Vlakancic (1999), Webb et al. (2000), Zdaniuk et al. (2008), Esen et al. (1994), Mukkamala and Sundaresan (2009), Meyer and Olivier (2011a, b), Tam et al. (2011), and Tam and Ghajar (1997, 2006) reported on the heat transfer and pressure drop of microfin tube in different flow region.

Zhang et al. (2012) carried out an experimental investigation on convective condensation of R410A in microfin tubes with same outer diameter and helix angle. They studied the effect of mass flux surface area on the heat transfer

Table 7.3 Start and end of transition of friction and heat transfer data for plain and microfin tubes at x/D_i of 200 (Tam et al. 2012)

Tube, condition	Friction factor				Heat transfer			
	Re_{start}	C_f	Re_{end}	C_f	Re_{start}	$St\,Pr^{0.67}$	Re_{end}	$St\,Pr^{0.67}$
Plain, isothermal (square-edged)	2306	7.6×10^{-3}	3588	0.0102	–	–	–	–
Plain, heating (square-edged)	2300	7.2×10^{-3}	3941	0.0100	2298	1.8×10^{-3}	8357	4.1×10^{-3}
Plain, isothermal (re-entrant)	2032	9.0×10^{-3}	3031	0.0110	–	–	–	–
Plain, heating (re-entrant)	2001	7.6×10^{-3}	3039	0.0106	2001	2.0×10^{-3}	7919	4.1×10^{-3}
Microfin #1, isothermal (square-edged)	2675	8.4×10^{-3}	8800	0.0144	–	–	–	–
Microfin #1, heating (square-edged)	2764	7.3×10^{-3}	9156	0.0138	2751	1.7×10^{-3}	8963	9.4×10^{-3}
Microfin #1, isothermal (re-entrant)	2021	0.0104	8496	0.0143	–	–	–	–
Microfin #1, heating (re-entrant)	2167	9.2×10^{-3}	9027	0.0143	2145	1.9×10^{-3}	8014	9.2×10^{-3}
Microfin #2, isothermal (square-edged)	2284	9.9×10^{-3}	8359	0.0154	–	–	–	–
Microfin #2, heating (square-edged)	2390	7.9×10^{-3}	8354	0.0155	2402	1.8×10^{-3}	8956	0.0106
Microfin #2, isothermal (re-entrant)	1973	0.0114	8342	0.0154	–	–	–	–
Microfin #2, heating (re-entrant)	1997	9.8×10^{-3}	8278	0.0158	1946	2.2×10^{-3}	7791	0.0109
Microfin #3, isothermal (square-edged)	1962	0.0104	8302	0.0170	–	–	–	–
Microfin #3, heating (square-edged)	2250	9.6×10^{-3}	8050	0.0168	2144	2.0×10^{-3}	8051	0.0116
Microfin #3, isothermal (re-entrant)	1849	0.0119	7989	0.0167	–	–	–	–
Microfin #3, heating (re-entrant)	1954	0.0110	8106	0.0169	1903	2.3×10^{-3}	7170	0.0124

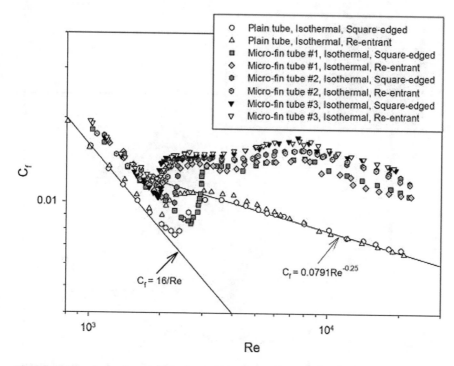

Fig. 7.5 Friction factor characteristics for the plain tube and the microfin tube at x/D_i of 200 under isothermal boundary conditions (Tam et al. 2012)

enhancement and interfacial turbulence. It was found that heat transfer rate increased due to increase in surface area at higher mass fluxes. Table 7.4 shows the dimensions of the tested tubes. Table 7.5 presents the seven existing correlations for calculating the pressure drop. Table 7.6 shows the values of seven frictional pressure drop and four values of heat transfer coefficient for all tubes from the existing correlations.

Table 7.7 describes the four existing correlations for condensation heat transfer coefficient. Figure 7.7 shows the variation of frictional pressure drop versus mass flux. Tube 1 and tube 5 presented maximum and minimum frictional loss, respectively, at maximum value of mass flux. Tube 4 has been found to have the best heat transfer performance due to its largest condensation heat transfer coefficient and relatively low pressure drop. Figure 7.8 shows the variation of heat transfer coefficient with respect to mass flux. Thome (2004), Yang (1999), Schlager et al. (1990), Kim et al. (2009), Akhavan-Behabadi et al. (2007), Kwon et al. (2000), Sapali and Patil (2010), Olivier et al. (2007), Cavallini et al. (1999, 2009), Choi et al. (2001), Kedzierski and Goncalves (1999), Huang et al. (2010), Wu and Li (2011) and Li and Wu (2010a, b, c) reported on the characteristics of microfin tubes that helped in the improvement of heat enhancement.

Brognaux et al. (1997) measured the effect of Prandtl number ($0.70 \leq Pr \leq 7.85$) for turbulent flow of water in microfin tubes. They have correlated their data by

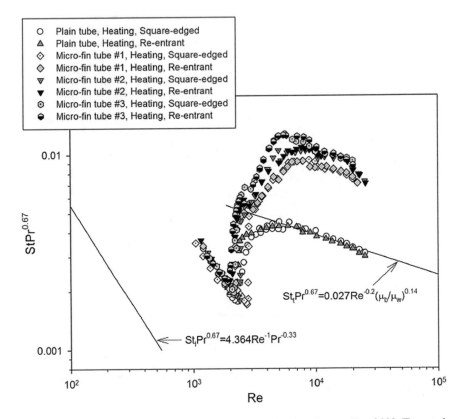

Fig. 7.6 Heat transfer characteristics for the plain and microfin tubes at x/D_i of 200 (Tam et al. 2012)

Table 7.4 Geometries of the tested tubes (Zhang et al. 2012)

Tube no.	d_0 (mm)	d_{fr} (mm)	η_s (−)	α (°)	β (°)	e (mm)	A_{ai}/A_{fr} (−)	e/p_f (mm)	d_h (mm)
Tube 1	5.0	4.6	38	40	18	0.16	1.61	0.43	2.77
Tube 2	5.0	4.6	38	25	18	0.16	1.71	0.44	2.65
Tube 3	5.0	4.54	36	25	18	0.12	1.51	0.32	2.97
Tube 4	5.0	4.54	60	25	18	0.12	1.85	0.53	2.40
Tube 5	5.0	4.6	52	20	18	0.10	1.63	0.38	2.77

$$Nu_{di,m} = C Re_{di,m}^{0.81} Pr_l^{0.55} \tag{7.6}$$

where $C = 0.02271 + 3.72\text{E} - 05\alpha - 9.337\text{E} - 7\alpha^3$, and α is the helix angle. Narayanamurthy (1999) extended the work of Brognaux et al. (1997) and covered Reynolds number range of 5000–70,000, and he developed multiple regression correlations for j and f factors. More advanced works on internal fin geometries

Table 7.5 Description of seven existing correlations for condensation pressure drop (Zhang et al. 2012)

Authors	Equations
Choi et al. (2001) (microfin tubes)	$\Delta p = \Delta p_f + \Delta p_a = G^2\left[\frac{f_c L(\nu_{out}+\nu_{in})}{d_h} + (\nu_{out}-\nu_{in})\right]$ $f_c = 0.00506 Re_{h,LO}^{-0.0951} K_f^{0.1554}, Re_{h,LO}=\frac{Gd_h}{\mu_1}, K_f = \Delta x h_v/(gL)$
Kedzierski and Goncalves (1999) (microfin tubes)	$f_c = 0.002275 + 0.00933 exp^{[c/(-0.003d_{fr})]} Re_{h,LO}^{-1/(4.16+532c/d_{tr})} K_f^{0.211}$ $Re_{h,LO} = \frac{Gd_h}{\mu_1}$
Haraguchi et al. (1993) (microfin tubes)	$(dp/dz)_1 = \Phi_v^2(dp/dz)_v = \Phi_v^2 2 f_v (Gx)^2/(\rho_v d_{fr})$ $\Phi_v = 1.1 + 1.3\{GX_{tt}/[gd_m\rho_v(\rho_1-\rho_v)]^{0.5}\}^{0.35}$
Beattie and Whalley (1982)	$\left(\frac{dp}{dz}\right)_f = \frac{2f_{tp}G^2}{d_{fr}\rho_{tp}}, Re_{tp}=\frac{Gd_{fr}}{\mu_{tp}}$ $\mu_{tp} = \mu_1 - 2.5\mu_1\left[\frac{x\rho_1}{x\rho_1+(1-x)\rho_v}\right]^2 + \left[\frac{x\rho_1(1.5\mu_1+\mu_v)}{x\rho_1+(1-x)\rho_v}\right]$
Friedel (1979)	$\left(\frac{dp}{dz}\right)_f = \left(\frac{dp}{dz}\right)_{LO}\Phi_{LO}^2, \left(\frac{dp}{dz}\right)_{LO} = f_{LO}\frac{2G^2}{d_{fr}\rho_1}, Re_{LO}=\frac{Gd_{fr}}{\mu_1},$ $Fr_{tp} = \frac{G^2}{gd_{fr}\rho_{tp}^2}, We_{tp}=\frac{G^2d_{fr}}{\sigma\rho_{tp}}$ $\Phi_{LO}^2 = (1-x)^2 + x^2\frac{\rho_1 f_{vo}}{\rho_v f_{LO}} + \frac{3.24x^{0.78}(1-x)^{0.224}(\rho_1/\rho_v)^{0.91}}{Fr_{tp}^{0.045}We_{tp}^{0.035}(\mu_v/\mu_1)^{-0.19}(1-\mu_v/\mu_1)^{-0.7}}$
Muller-Steinhagen and Heek (1986)	$\left(\frac{dp}{dz}\right)_f = \left\{\left(\frac{dp}{dz}\right)_{LO} + 2\left[\left(\frac{dp}{dz}\right)_{VO} - \left(\frac{dp}{dz}\right)_{LO}\right]x\right\}(1-x)^{1/3} + \left(\frac{dp}{dz}\right)_{VO}x^3$ $Re_{VO} = \frac{Gd_{fr}}{\mu_v}, \left(\frac{dp}{dz}\right)_{VO} = f_{VO}\frac{2G^2}{d_{fr}\rho_v}$
Gronnerud (1979)	$\left(\frac{dp}{dz}\right)_f = \left(\frac{dp}{dz}\right)_{LO}\Phi_{LO}^2, (\Phi)_{LO} = 1 + \left(\frac{dp}{dz}\right)_{Fr}\left[\frac{\rho_1/\rho_v}{(\mu_1/\mu_v)^{0.25}} - 1\right]$ $\left(\frac{dp}{dz}\right)_{Fr} = f_{Fr}\left[x + 4(x^{1.8} - x^{10}f_{Fr}^{0.5})\right], Fr_{LO}=\frac{G^2}{gd_{fr}\rho_1^2}$ If $Fr_{LO} \geq 1.0, f_{Fr} = 1.0$; or if $Fr_{LO} < 1.0, f_{Fr} = Fr_L^{0.3} + 0.0055\left(\ln\frac{1}{Fr_L}\right)^2$

Table 7.6 Evaluation of the seven pressure drop and the four heat transfer correlations (Zhang et al. 2012)

Authors	Tube 1	Tube 2	Tube 3	Tube 4	Tube 5
Frictional pressure drop					
Choi et al. (2001)	41.5	21.5	30.8	12.5	20.9
Kedzierski and Goncalves (1999)	31.5	9.9	25.9	7.7	23.7
Haraguchi et al. (1993)	51.0	35.1	36.0	38.2	36.5
Beattie and Whalley (1982)	13.9	11.2	7.8	7.3	6.6
Friedel (1979)	26.5	7.1	6.7	11.1	7.6
Muller-Steinhagen and Heck (1986)	26.4	7.8	7.1	11.0	8.8
Gronnerud (1979)	17.2	8.9	7.4	7.4	7.3
Heat-transfer coefficient					
Cavallini et al. (2009)	39.2	27.1	35.8	74.6	57.6
Kedzierski and Goncalves (1999)	95.8	72.1	85.2	116.3	85.1
Moser et al. (1998)	23.9	25.2	23.9	23.5	31.5
Shah (1979)	53.3	39.3	64.6	35.4	54.6

Table 7.7 Description of four correlation heat transfer coefficients (Zhang et al. 2012)

Authors	Equations
Cavallmi et al. (2009) (microfin tube)	$Nu_{ft} = h_{ft}d_t/k_1 = 0.05Re_{eq}^{0.8}Pr_1^{1/3}Rx^{2.0}(Bo_{Webb}Fr_{VO})^{-0.26}$ $Re_{eq} = 4\dot{m}\left[(1-x)+x(\rho_1/\rho_v)^{0.5}\right]/(\pi d_{ft}\mu_1)$ $Rx = \{[2en_s(1-\sin(\alpha/2))]/[\pi d_{fr}\cos(\alpha/2)]+1\}/\cos\beta$ $Bo_{Webb} = g\rho_1 e\pi d_{fr}/(8\sigma n_s), Fr_{VO} = u_{VO}^2/(gd_{fr})$
Kedzierski and Goncalves (1999) (microfin tube)	$Nu_{ai} = h_{ai}d_h/k_1 = 4.94Re_h^{0.235}Pr_1^{0.308}(p_{red})^{-1.16x^2}(-\log_{10}p_{red})^{-0.887x^2}Sv^{2.708x}$ $Re_h = Gd_h/\mu_1, Sv = (v_v - v_1)/v, v = xv_v + (1-x)v_1, Pr_1 = c_{p1}\mu_1/k_1$
Moser et al. (1998)	$Nu_{ai} = h_{ai}d_{fr}/k_1 = \dfrac{0.0994^{c1}Re_1^{c2}Re_{eq}^{1+0.875c1}Pr_1^{0.815}}{\left(1.58\ln Re_{eq}-3.28\right)\left(2.58\ln Re_{eq}+13.7Pr_1^{2/3}-19.1\right)}$ $c_1 = 0.126Pr_1^{-0.448}, c_2 = -0.113Pr_1^{-0.563}, R+ = 0.0994Re_{eq}^{7/8}, Re_{eq} = \Phi_{LO}^{8/7}Re_{LO}$
Shah (1979)	$\dfrac{h_{fr}}{h_{LO}} = (1-x)^{0.8} + \dfrac{3.8x^{0.76}(1-x)^{0.04}}{p_{red}^{0.38}}, h_{LO} = 0.023Re_{LO}^{0.8}Pr_1^{0.3}\cdot d_{fr}/k_1$

Fig. 7.7 Condensation frictional pressure drop vs. mass flux (Zhang et al. 2012)

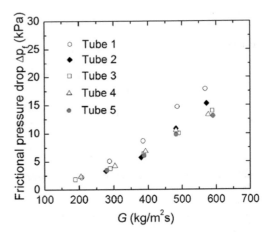

Fig. 7.8 Condensation heat transfer coefficient vs. mass flux (Zhang et al. 2012)

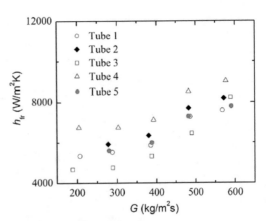

Fig. 7.9 (a) Axial fins on external surface, (b) axial fins used in double-pipe heat exchanger, (c) axial fins with multitubes, (d) cut-and-twist axial fins, (e) offset strip fins (Webb and Kim 2005)

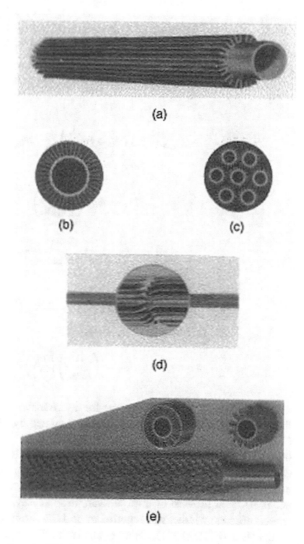

(a)

(b)   (c)

(d)

(e)

have been reported by Wang et al. (1996), Moser et al. (1998) and Bhatia and Webb (2001). Figure 7.9 shows axial fins (finned annuli), which are used for a double-pipe heat exchanger or for axial flow on the outer surface of a tube bundle.

DeLorenzo and Anderson (1945), Gunter and Shaw (1942), Taborek (1997), Braga and Saboya (1999) and Edwards et al. (1988) have reported several works with finned annuli and the correlations are given by

$$Nu_{Dh} = Nu_p \left[0.86 \left(\frac{D_o}{D_i} \right) \right] \tag{7.7}$$

$$Nu_{Dh} = \left[Nu_L^z + Nu_x^z\right]^{1/z}\left(\frac{\mu_b}{\mu_w}\right)^n \tag{7.8}$$

$$Nu_L = \left[Nu_\infty^3 + Nu_{L,a}^3\right]^{1/3} \tag{7.9}$$

$$Nu_{L,a} = 2.1\left(Re_{Dh}Pr\frac{D_h}{L}\right)^{1/3}, \quad Nu_\infty = 4.12 \tag{7.10}$$

$$Nu_x = Nu_{tr}\left(\frac{Re_{Dh}}{15,000}\right)^{1.25} \tag{7.11}$$

$$f = \left[f_{tub}^3 + f_{lam}^3\right]^{1/3} \tag{7.12}$$

$$f_{tub} = \left[1.58\ln(Re_{Dh}) - 3.28\right]^{-2}\left(\frac{\mu_b}{\mu_w}\right)^n \tag{7.13}$$

$$f_{lam} = \left(\frac{16}{Re_{Dh}}\right)\left(\frac{\mu_b}{\mu_w}\right)^n \tag{7.14}$$

The performance of microfin tube for heat transfer augmentation of single-phase flow in the transition regime has been investigated by Mukkamala and Sundaresan (2009). They used hot water as the working fluid in the double-pipe microfin tube heat exchanger. The heat transfer enhancement ratio, isothermal enhancement index and the efficiency index for microfin tubes have been presented in Figs. 7.10, 7.11 and 7.12, respectively. They concluded that the augmentation in heat transfer was about 107% at the cost of 20% pressure drop at Reynolds number of 12,621. The overall efficiency index was reported to be 1.77 corresponding to $Re = 12,621$.

Eckels and Pate (1991), Chiou et al. (1995), Wang et al. (1996), Brognaux et al. (1997), Copetti et al. (2004), Han and Lee (2005), Al-Fahed et al. (1999) and Wang and Rose (2004) have worked with microfin tubes.

Wang et al. (2019) carried out an experimental comparison of the heat transfer of supercritical R134a in a microfin tube and a smooth tube for mass fluxes of 100–700 kg/m²s, heat flux of 10–70 Kw/m², and pressure drop of 4.26–5 MPa. They investigated the effect of heat fluxes and buoyancy on the wall of fin tube and heat transfer coefficient. Heat transfer coefficient of top wall of smooth tube had reduced more than microfin tube due to buoyancy effect. They observed from the experimental results that heat transfer coefficient in the microfin tube was 1.68 times on the top and 1.59 times on the bottom of smooth tube. Figure 7.13 shows variation of wall temperature and heat transfer coefficient of top and bottom wall of both the

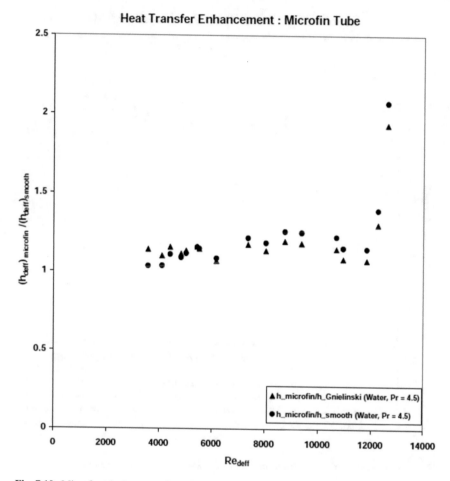

Fig. 7.10 Microfin tube heat transfer enhancement (Mukkamala and Sundaresan 2009)

types of tubes with enthalpy at constant pressure, constant inlet temperature, mass flux (G) and heat flux (q). Thus, it was observed that wall temperature difference of the microfin tube had smaller change than smooth tube as heat flux was increased. Higahiiue et al. (2007), Kuwahara et al. (2012), Lee et al. (2013), Liu et al. (2017) and Kim and Kim (2010) studied the effect of microfin tube in heat exchanger.

Li et al. (2007) experimentally investigated the pressure drop and heat transfer characteristics in a microfin tube. They used water and oil in the testing tube with a huge range of Prandtl number and Reynolds number. They presented Fig. 7.14 showing the microfin tube and its geometrical parameters. Both the smooth and the microfin tubes are made up of copper material. They varied the Prandtl number from 3.2 to 5.8 for water and 80 to 220 for oil. Similarly, for experimentation, they varied Reynolds number from 2500 to 90,000 for the case of water and 2500 to 12,000 for the working fluid oil. The analysis revealed that the heat transfer

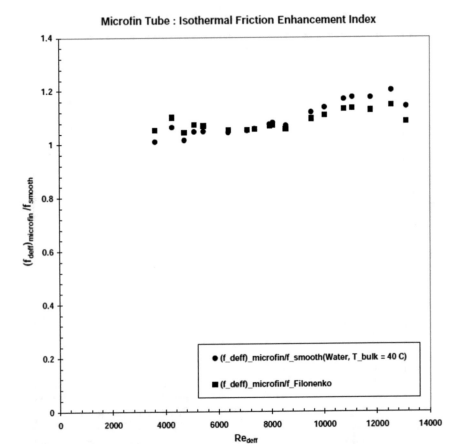

Fig. 7.11 Isothermal friction enhancement index (Mukkamala and Sundaresan 2009)

enhancement was not profound with water at low Prandtl number until Reynolds number 10,000 (defined as critical Reynolds number) as Nusselt number for both microfin and smooth were approximately similar.

It increases rapidly above Reynolds number 10,000 and became twice at 30,000. Also, critical Reynolds number changed with the Prandtl number, and it became 6000 for higher Prandtl number. They observed the friction factor of the microfin tube and concluded that when Reynolds number was less than 10,000, the friction factor behaviour was same as that of smooth tube but when it was more than 10,000, the friction factor increased and reached up to 40–50% greater than the smooth tube. They again follow the same trend as that of smooth tube beyond 30,000. Both the lines became parallel afterwards when we crossed the upper limit of Reynolds Number 90,000. This is presented in Fig. 7.15. They concluded that for low Reynolds number, the microfin resides inside viscous sub-layer and thus acts like a smooth tube. However, viscous sub-layer decreased with the increased Reynolds

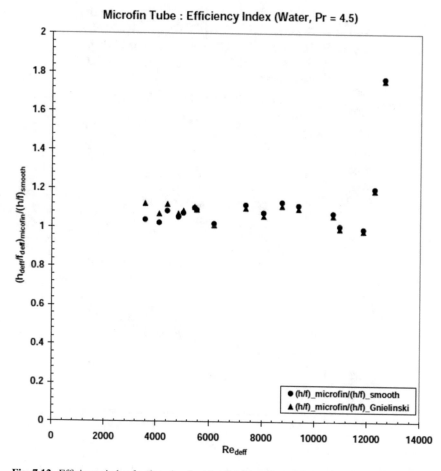

Fig. 7.12 Efficiency index for the microfin tube (Mukkamala and Sundaresan 2009)

number, and the microfin started to show its enhancement effect after critical Reynolds number.

Nuntaphan et al. (2005) studied and investigated the air-side performance for heat transfer and friction characteristics of cross-flow heat exchangers consisting of crimped spiral fins under dehumidification condition. Many researchers worked on this topic such as Briggs and Young (1963), Robinson and Briggs (1966), Rabas et al. (1981), Nuntaphan and Kiatsiriroat (2003) worked on air-side performance calculation. However, there is no data available for crimped spiral fins under dehumidification. They presented crimped spiral fins in Fig. 7.16, and the geometrical aspects of cross-flow heat exchanger are presented in Table 7.8. Ten crimped spiral fin exchanger having different geometric parameters were tested. They developed the equation of Colburn j factor as

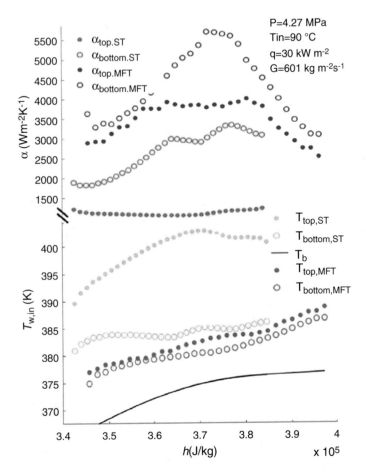

Fig. 7.13 Wall temperature and heat transfer coefficient variations (Wang et al. 2019)

$$j = 0.0208 Re_D^m \left(\frac{d_o}{S_t}\right)^{-2.5950} \left(\frac{f_t}{f_s}\right)^{0.7905} \left(\frac{S_1}{S_t}\right)^{0.2391} \left(\frac{d_o}{d_f}\right)^{0.2761} \qquad (7.15)$$

taking into consideration the effect of geometric parameter on heat transfer performance. Also, they plotted heat transfer coefficient and the pressure drop with frontal velocity and concluded that wet surface heat transfer coefficient was lesser than the corresponding dry surface coefficient. They concluded that larger tube diameter performed not good and had lower heat transfer coefficient than the smaller one. It was also observed that it increased pressure drop penalty. They found that fin height has no influence under wet condition. They concluded that fin spacing has negligible effect on heat transfer performance but increasing fin spacing leads towards lower heat transfer coefficient. Results revealed that higher heat transfer coefficient can be obtained by lower transverse pitch.

Fig. 7.14 (**a**) Microfin tube and (**b**) fin geometry parameters (Li et al. 2007)

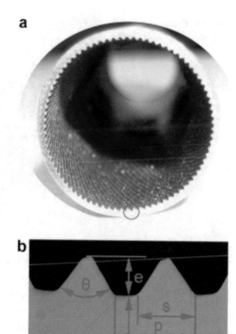

Fig. 7.15 Friction factors in the microfin tube (Li et al. 2007)

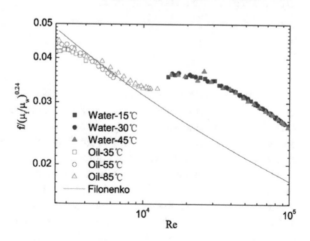

Gharebaghi and Sezai (2007) enhanced the thermal energy of storage unit by inserting fin array system into the storage device. Thermal energy storage unit was filled with phase change material (PCM). Aluminium fins were attached to the storage walls through which heat was transferred. Paraffin wax was used as a

Fig. 7.16 Crimped spiral fins (Nuntaphan et al. 2005)

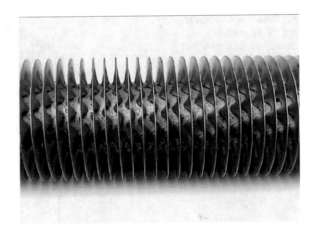

Table 7.8 Geometric dimensions of cross-flow heat exchanger (Nuntaphan et al. 2005)

Sample	d_o (mm)	d_i (mm)	f_s (mm)	f_h (mm)	f_t (mm)	S_t (mm)	S_1 (mm)	n_r	n_t	Arrangement
1	17.3	13.3	3.85	10.0	0.4	50.0	43.3	4	9	Staggered
2	21.7	16.5	6.10	10.0	0.4	72.0	36.0	4	6	Staggered
3	21.7	16.5	3.85	10.0	0.4	72.0	36.0	4	6	Staggered
4	21.7	16.5	2.85	10.0	0.4	72.0	36.0	4	6	Staggered
5	21.7	16.5	6.10	10.0	0.4	84.0	24.2	4	5	Staggered
6	21.7	16.5	3.85	10.0	0.4	84.0	24.2	4	5	Staggered
7	21.7	16.5	2.85	10.0	0.4	84.0	24.2	4	5	Staggered
8	21.7	16.5	3.85	10.0	0.4	55.6	48.2	4	8	Staggered
9	21.7	16.5	3.85	15.0	0.4	55.6	48.2	4	8	Staggered
10	27.2	21.6	3.85	10.0	0.4	50.0	43.3	4	9	Staggered

phase change material and stored between the fins. They developed the mathematical model to solve the melting problem in a two-dimensional domain. Non-uniform grids for different fins and PCM layer thickness were analysed by transient simulation at constant ratio of PCM layer to fin thickness. Table 7.9 shows the thermophysical parameters of PCM and aluminium.

They observed that the time required for complete melting for both horizontal and vertical modules can be reduced by decreasing the fin spacing. Minimum time for melting was desirable for designing large-capacity storage units of small temperature difference. It was also observed that the Nusselt number was higher for vertical module arrangement compared to horizontal module for all fin spacing and temperature difference values. They found that heat flux was increased by inserting the fin system into the modules at larger value of temperature difference. Total storage capacity of thermal storage device can be enhanced by decreasing the inter-fin distance and module thickness. Lacroix and Benmadda (1997), Vakilaltojjar and Saman (2001), Zalba et al. (2004), Saman et al. (2005), Voller (1990) and

Table 7.9 Thermophysical properties of PCM and Al (Gharebaghi and Sezai 2012)

Property	Typical values	
	RT27	Aluminium
k (W/m K)	0.2	202.4
ρ (kg/m^3)	750	2719
C_p (J/kg K)	1800	871
β (1/K)	0.001	–
ν (mm^2/s)	4.5	–
L (kJ/kg)	179	–
Melting point (°C)	28	–
α (m^2/s)	1.48×10^{-7}	8.54×10^{-5}

Gharebaghi (2007) investigated the phase change process of PCM in a thermal energy storage system with fins.

Agarwal (2016) numerically simulated melting process of phase change materials (PCM) paraffin wax in a horizontal cylindrical annulus. Regin et al. (2008) studied phase change materials and revealed that PCM is very interesting because of its capacity to store/release thermal energy through solid–liquid phase change process. Agarwal (2016) used longitudinal fins to enhance heat transfer. He investigated three conditions: (1) without fins, (2) with four fins 90° apart and (3) with eight equally spaced fins.

Many researchers Khodadadi and Zhang (2001), Zalba et al. (2004), Agarwal and Sarviya (2016), Hosseini et al. (2012) and Seeniraj et al. 2002) examined the melting of PCM and investigated heat transfer characteristic of internally finned PCM. Typically, PCM possess low thermal conductivity. Thus, different geometrical shaped metal fins were inserted to improve heat transfer. Rozenfeld et al. (2015), Ogoh and Groulx (2012), Al-Abidi et al. (2013), Liu and Groulx (2014), Murray and Groulx (2014), Sciacovelli et al. (2014), Sharifi et al. (2011), Mat et al. (2013), Li and Liu (2013) and Manglik and Jog (2016) investigated the heat transfer enhancement due to the addition of fins conjugated with phase changing materials.

Agarwal (2016) presented Fig. 7.17 for schematic view of computational domain with and without fin arrangement. He approximated it in two-dimensional model and solved non-linear partial differential equation. He utilized semi-implicit method for pressure-linked equations (SIMPLE) algorithm for better solution. Figure 7.18 reveals the melt fraction contours which are concentric to inner cylinder initially. After a lapse of 500 s, melting front seemed to be irregular due to natural convection-dominated conduction, and after 1000 s, the melting front was found to be wider in the upper part compared to that in the lower part. It was found that there was a rapid rise in the melt fraction in the beginning and that it became slower with time because the conduction mode overpowers the convection mode.

The results of four fins are presented in Fig. 7.19. He plotted transient variation and displayed the melt fraction with time in which the finned tube took much less time for complete melting of PCM. Similar trends were observed with cylindrical annulus consisting of eight fins where the time taken for complete melting was very less in comparison to the other two. The transient variation graph was steeper in case

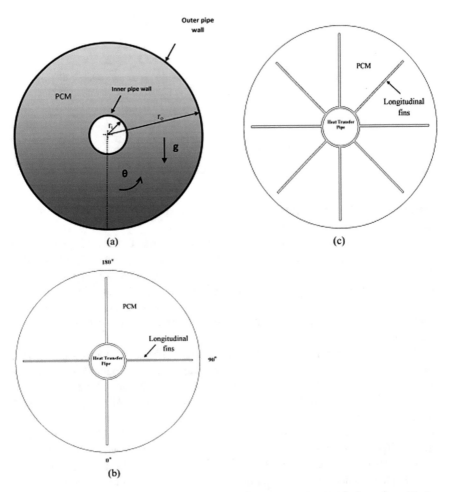

Fig. 7.17 Computational domains of the plain annulus, the finned cylindrical annulus with four fins, and the finned cylindrical annulus with eight fins (Agarwal 2016)

of eight fins, followed by that in case of four fins and then in smooth tube. All three cases have been shown in Fig. 7.20. Also, he presented the effectiveness of fins through fin effectiveness versus time graph where eight fins including cylindrical annulus performed better initially and after 1500 s they showed comparable effectiveness, which has been shown in Fig. 7.21. From experimental data, he recommended longitudinal eight fins for double-pipe heat storage with PCM.

Togun et al. (2016) numerically examined the three-dimensional heat transfer and flow separation of Al_2O_3/nanofluid flow in concentric annular pipe. It happened due to the variation in pressure gradient developed by change of annular cross-sectional area. Hussein et al. (2015), Togun et al. (2013), Safaei et al. (2014) and Hussein et al. (2013) investigated this phenomenon. Many researchers like Chieng and Launder

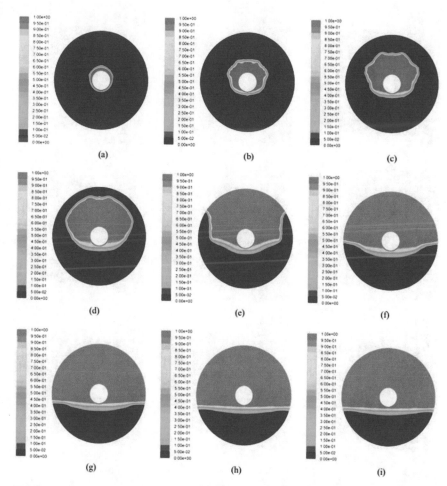

Fig. 7.18 Melt fraction contours in the PCM at different instants of time for a plain cylindrical annulus. (**a**) Melt contour at 100 s. (**b**) Melt contour at 500 s. (**c**) Melt contour at 1000 s. (**d**) Melt contour at 1500 s. (**e**) Melt contour at 2000 s. (**f**) Melt contour at 2500 s. (**g**) Melt contour at 3000 s. (**h**) Melt contour at 3500 s. (**i**) Melt contour at 4000 s (Agarwal 2016)

(1980), Chung and Jia (1995), Hsieh and Chang (1996), Launder and Shirma (1974), Lam and Bremhorst (1981), Nagano and Hishidu (1987), Myong (1990), Nagano and Tagawa (1990), Yang (1999), Abe et al. (1994) and Chang et al. (1995) numerically investigated the heat transfer for shear-free layer flow with smaller pressure gradient and used standard k-ε model because it provides good results.

The objective of their study was to numerically investigate the turbulent convective heat transfer in sudden enlargement of annular passage and examine the effects of volume fraction of nanoparticles, Reynolds number and expansion ratio. They considered parameters for heat transfer analysis. They used pure water or Al_2O_3/water with different concentrations under no slip condition. Boundary conditions,

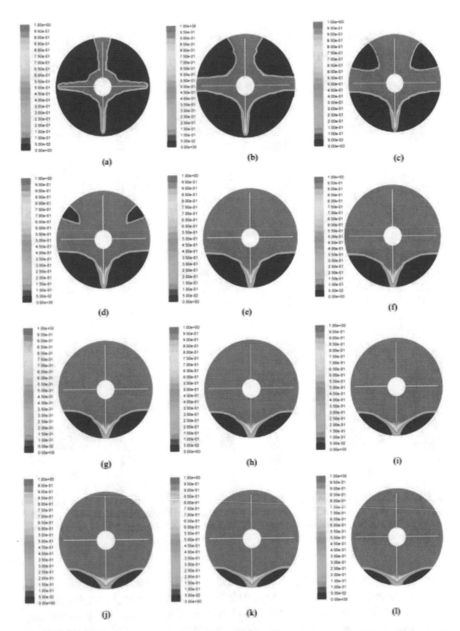

Fig. 7.19 Melt fraction contours in the PCM at different instants of time for a finned cylindrical annulus (four fins). (**a**) Melt contour at 200 s. (**b**) Melt contour at 400 s. (**c**) Melt contour at 600 s. (**d**) Melt contour at 800 s. (**e**) Melt contour at 1000 s. (**f**) Melt contour at 1200 s. (**g**) Melt contour at 1400 s. (**h**) Melt contour at 1600 s. (**i**) Melt contour at 1800 s. (**j**) Melt contour at 1800 s. (**k**) Melt contour at 2000 s. (**l**) Melt contour at 2200 s (Agarwal 2016)

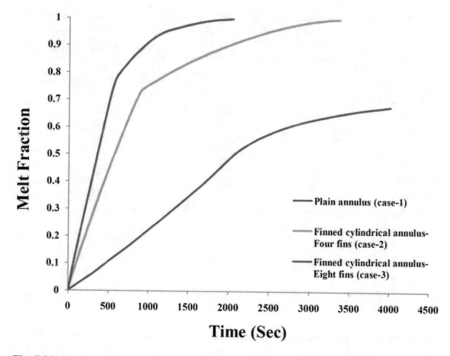

Fig. 7.20 Transient variations of the melt fraction for the plain annulus, the finned cylindrical annulus with four fins, and the finned cylindrical annulus with eight fins (Agarwal 2016)

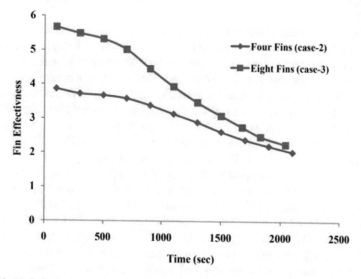

Fig. 7.21 Variations of fin effectiveness for four and eight fins (Agarwal 2016)

Table 7.10 Boundary conditions, thermophysical properties of Al_2O_3 and water and grid independent for pure water at ER = 2, q = 4000 W/m^2 and Re = 20,000 (Togun et al. 2016)

Boundary conditions				20,000, 30,000, 40,000, 50,000
Expansion ratio (ER)				1.25, 1.67 and 2
Heat flux (W/m^2)				4000, 8000, 12,000, 16,000
Volume fraction of Al_2O_3				0.5, 1, 1.5, 2
Thermophysical properties of the Al_2O_3 nanoparticles and water at T = 300 K				
Thermophysical properties		Al_2O_3		Water
$\square$ (kg/m^3)		3600		996.5
c_p (J/kg k)		765		4181
K (W/m k)		36		0.613
M (Ns/m^2)		–		1E – 03
Grid-independent for pure water at ER = 2, q = 4000 W/m^2 and Re = 20,000				
Size of mesh	X = 30, Y = 30 and Z = 300	X = 20, Y = 20 and Z = 500	X = 20, Y = 20 and Z = 750	X = 30, Y = 30 and Z = 750	X = 20, Y = 20 and Z = 1000
Grid number	1	2	3	4	5
h_{ave}	1233.43027	1246.70394	1247.764473	1248.151331	1248.237669

thermophysical properties of Al_2O_3 nanoparticles and water at temperature 300 K and grid-independent for pure water at Reynolds number 20,000 are presented in Table 7.10. They simulated four volume fraction values of Al_2O_3 and water (Φ = 0.5%, 1%, 1.5% and 2%) with the uniform heat flux ranges from 4000 to 16,000 W/m^2 and at Reynolds number range of 20,000 $\leq Re \leq$ 50,000. They observed the effect of expansion ratio and concluded that heat transfer coefficient increased up to maximum and then decreased to a constant value as shown in Fig. 7.22.

Other results revealed that pure water heat transfer coefficients were lower than that of Al_2O_3 nanofluids. They studied the effect of increased Reynolds number and found that heat transfer coefficient increased but did not alter the location at which highest heat transfer coefficient was obtained. The increased concentration of Al_2O_3 nanofluid increased the heat transfer coefficient at the outer cylinder surface. They presented contours for kinetic energy in Fig. 7.23 and accounted that kinetic energy simultaneously increased with increase of expansion ratio. They also presented Fig. 7.24 where streamlines for Reynolds number Re = 20,000, 30,000, 40,000 and 50,000 with expansion ratio = 2 were plotted respectively.

Raj et al. (2015) investigated the heat transfer and pressure drop in double-pipe heat exchanger. The testing was conducted in laminar-transient-turbulent flow regime with water and ethylene glycol as working fluids. Wen-Tao Ji et al. (2012), Ayub et al. (2006), Ooi et al. (2004), Park and Jung (2008), Al-Fahed et al. (1999), Ito and Kimura (1979) and Koyama et al. (1993) worked on doubly enhanced tube and horizontal microfin tube for heat transfer enhancement using water and ethylene glycol. They presented the dimensions of heat exchanger tube in Table 7.11. They experimented with the enhanced tube and plotted the variation of heat transfer

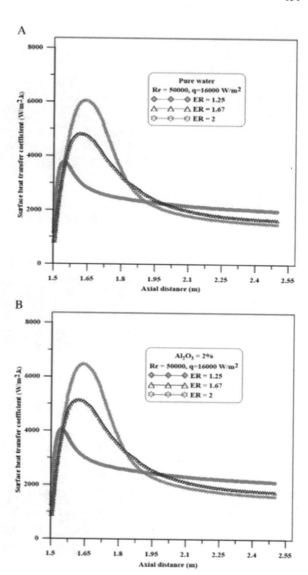

Fig. 7.22 Variations of surface heat transfer coefficient at different expansion ratios and $Re = 50,000$ for (**a**) pure water and (**b**) 2% Al_2O_3 nanofluid (Togun et al. 2016)

coefficient with Reynolds number Re_{deff} in Fig. 7.25 under laminar-transition-turbulent regime and only laminar region. The heat transfer enhancement was 34% with water in Turbo-C tube at $Re_{deff} = 18,187$, whereas in Turbo-CDI tube, it was 23% at $Re_{deff} = 17,639$. This result shows that Turbo-C achieved higher efficiency ratios due to smaller effective diameter and it provokes higher flow velocity with higher heat transfer and greater wall shear.

They also found that in laminar regime, enhancement ratio has no significance as it was 6% and 4% with Turbo-C and Turbo-CDI tubes, respectively. The overall enhancement ratio was defined by $(UA_{eff})_{enhanced} / (UA_{eff})_{smooth}$. They obtained

Fig. 7.23 The counter of turbulent kinetic energy for 2% Al$_2$O$_3$ nanofluid and Reynolds number of 50,000; (**a**) ER = 1.25, (**b**) ER = 1.67 and (**c**) ER = 2 (Togun et al. 2016)

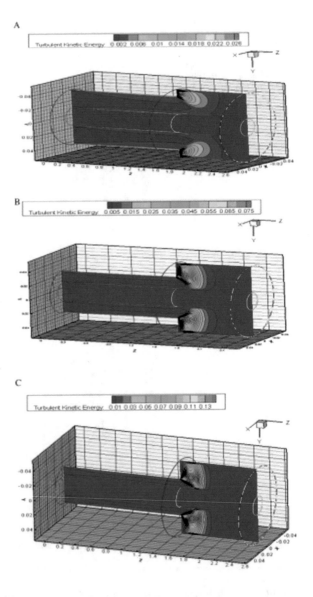

116% heat transfer enhancement with water in Turbo-C tube. They recommended that these tubes should be applied in fully developed rough flow regime at $Re_{deff} > Re_{critical}$. From the plotted data, they found negligible heat transfer enhancement in laminar regime. They examined the friction factor and pressure drop in doubly enhanced tube with water and ethylene glycol and found that tube-side friction factor was approximately identical to smooth tube friction factor. It confirms that enhanced tube is ineffective in laminar regime, and compound heat transfer technique should be employed under laminar regime. They concluded that doubly

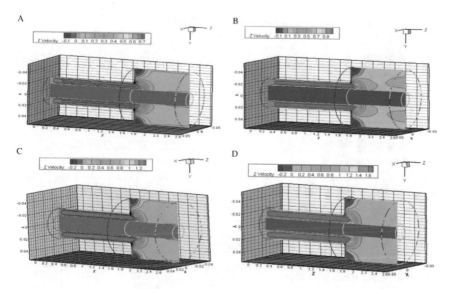

Fig. 7.24 The streamline of velocity at ER = 2 and ϕ = 2% for (**a**) Re = 20,000; (**b**) Re = 30,000; (**c**) Re = 40,000 and (**d**) Re = 50,000 (Togun et al. 2016)

enhanced tube produce best results when similar thermal conductivity fluids are flowing in the tube and annulus. They found that 54% reduction in pump duty for an indicated heat duty and it suggested that these tubes were best fitted in turbulent flow regime.

Yu et al. (1999) carried out an experimental study on pressure drop and heat transfer characteristics of tube with internal wave-like longitudinal fins in the entrance and fully developed regions. The test tube had double-pipe annulus structure and wave-like fin inserted in this annulus. They carried out experiments in two cases: one with inner tube blocked (no air entered through it) and the second with inner tube unblocked. The outer tube was heated electrically. They developed correlations for Nusselt number and friction factor in the fully developed region. They measured the local and average heat transfer coefficient and pressure drop in the Reynolds number range of 900–3500. It was observed that the highest heat transfer enhancement was obtained in case of blocked tube. Figure 7.26 shows the cross-section view of unblocked and blocked tube.

Figure 7.27 illustrates that the friction factor decreased with increasing value of Reynolds number in both the cases but at same Reynolds number, pressure dropped more in case of blocked tube. Figure 7.28 depicts the variation of Nusselt numbers with Reynolds numbers, and it was cleared that at the same Reynolds number, heat transfer was higher in case of blocked tube. Experimental results also showed that the thermal entrance length depended on Reynolds number. Kelkar and Patankar (1990), Fu et al. (1995) and Webb and Scott (1980) studied the performance of internally finned tube for heat exchanger application.

Table 7.11 Dimensions of heat exchangers (Raj et al. 2015)

Sl. no	Parameter	Smooth tube	Turbo-C [24]	Turbo-CDI [24]
1.	Cu, approach tube 2 m long at inlet, OD (mm)	19.05 (3/4″)	19.05 (3/4″)	19.05 (3/4″)
2.	Cu, approach tube ID	15.88 (5/8″)	15.88 (5/8″)	15.88 (5/8″)
3.	Cu, exit tube, 0.5 m long at exit, OD (mm)	19.05 (3/4″)	19.05 (3/4″)	19.05 (3/4″)
4.	Cu, exit tube ID (mm)	15.88 (5/8″)	15.88 (5/8″)	15.88 (5/8″)
5.	Cu, HX outer tube OD: D_o (mm)	25.4 (1″)	25.4 (1″)	25.4 (1″)
6.	Cu, HX outer tube ID: D_i (mm)	22.22 (7/8″)	22.22 (7/8″)	22.22 (7/8″)
7.	Cu, HX outer tube wall thickness: T_w (mm)	1.6	1.6	1.6
8.	Inner tube OD: d_o (mm) [24]	19.05	19.05	19.05
9.	Inner tube ID: d_i (mm) [24]	15.88 (5/8″)	15.29	15.54
10.	Inner tube wall thickness, t_w (mm) [24]	1.59	0.711	0.711
11.	Inner tube effective diameter, d_{eff} (mm) [24]	15.9	14.93	15.36
12.	Inner tube hydraulic diameter, d_h (mm) [24]	15.9	9.73	9.15
13.	Inner tube actual diameter, d_{actual} (mm) [24]	15.9	22.92	25.78
14.	Number of tube-side (internal) fins, $N_{f,i}$ (internal fins) [24]	0	30	35
15.	Internal fins spiral (helix) angle, α (helix angle) [24]	0	35°	40°
16.	Internal fin height, $e_{f,i}$ (mm) [24]	0	0.432	0.483
17.	External fin height, $e_{f,o}$ (mm) [24]	0	0.98	0.95
18.	External fin density (fins/inch (25.4 mm)) [24]	0	40	40
19.	$w_{f,avg}$ (average fin width, mm) [24]	0	0.66	0.262
20.	Inner tube perimeter, $P = \pi d_a$ (wetted perimeter, mm) [24]	49.95	72	80.99
21.	Inner tube A_{ff} (free flow area, mm^2) [24]	198.6	175.05	185.24
22.	Inner tube A_i (nominal area, mm^2) [24]	198.6	183.61	189.67
23.	Internal fin pitch/height, p/e_r [24]	0	5.36	3.4
24.	L (heat exchanger length, mm) [24]	2590	2590	2590
25.	Inner enhanced tube plain end length (mm) [24]	0	152.4	152.4
26.	Inner tube A_a (actual surface area, m^2/m) per metre	0.0499	0.072	0.081
27.	Inner tube area enhancement (EA) [24]	1.0	1.5	1.62

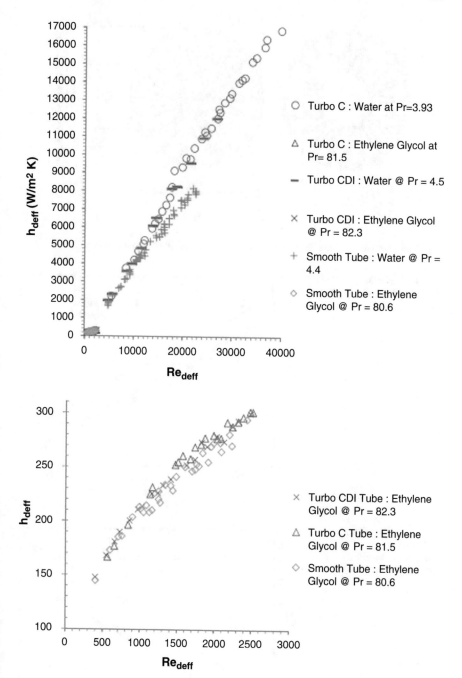

Fig. 7.25 Heat transfer coefficients in enhanced tubes. Heat transfer coefficients in enhanced tubes in laminar regime (Raj et al. 2015)

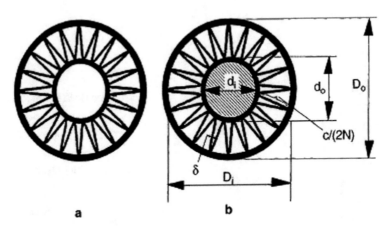

a   b

Fig. 7.26 The cross-section view of tube: (**a**) unblocked and (**b**) blocked (Yu et al. 1999)

Fig. 7.27 Friction
factor vs. Reynolds number
(Yu et al. 1999)

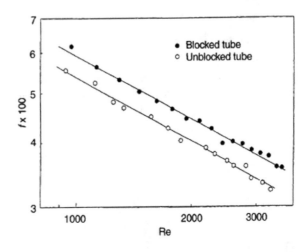

Fig. 7.28 Variation
of Nusselt
number vs. Reynolds
number (Yu et al. 1999)

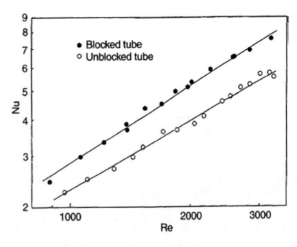

References

Abe K, Kondoh T, Nagano Y (1994) A new turbulence model for predicting fluid flow and heat transfer in separating and reattaching flows—I. Flow field calculations. Int J Heat Mass Transf 37(1):139–151

Agarwal A, Sarviya RM (2016) An experimental investigation of shell and tube latent heat storage for solar dryer using paraffin wax as heat storage material. Eng Sci Technol Int J 19(1):619–631

Akhavan-Behabadi MA, Kumar R, Mohseni SG (2007) Condensation heat transfer of R134a inside a microfin tube with different tube inclinations. Int J Heat Mass Transf 50:4864–4871

Al-Abidi AA, Mat S, Sopian K, Sulaiman MY, Mohammad A (2013) Internal and external fin heat transfer enhancement technique for latent heat thermal energy storage in triplex tube heat exchangers. Appl Therm Eng 53:147–156

Al-Fahed S, Chamra LM, Chakroun W (1999) Pressure drop and heat transfer comparison for both microfin tubes and twisted tape inserts in laminar flow. Exp Thermal Fluid Sci 18:323–333

Ayub ZH, Chyu MC, Ayub AH (2006) Case study: limited charge shell and tube ammonia spray evaporator with enhanced tubes. Appl Therm Eng 26(11–12):1334–1338

Beattie DH, Whalley PB (1982) Simple two-phase frictional pressure drop calculation method. Int J Multiphase Flow 8:83–87

Bhatia RS, Webb RL (2001) Numerical study of turbulent flow and heat transfer in microfin tubes—part 2, parametric study. J Enhanc Heat Transf 8:305–314

Braga CVM, Saboya FEM (1999) Turbulent heat transfer, pressure drop and fin efficiency in annular regions with continuous longitudinal rectangular fins. Exp Therm Fluid Sci 20:55–65

Briggs DE, Young EH (1963) Convection heat transfer and pressure drop of air flowing across triangular pitch banks of finned tubes. Chem Eng Prog Symp Ser 59(41):1–10

Brognaux LB, Webb RL, Chamra LM (1997) Single phase heat transfer in micro-fin tubes. Int J Heat Mass Transf 40(18):4345–4357

Cavallini A, Del Col D, Doretti L, Longo GA, Rossetto L (1999) A new computational procedure for heat transfer and pressure drop during refrigerant condensation inside enhanced tubes. J Enhanc Heat Transf 6:441–456

Cavallini A, Del Col D, Mancin S, Rossetto L (2009) Condensation of pure and near-azeotropic refrigerants in microfin tubes: a new computational procedure. Int J Refrig 32:162–174

Chang KC, Hsieh WD, Chen CS (1995) A modified low-Reynolds-number turbulence model applicable to recirculating flow in pipe expansion. Trans ASME I J Fluids Eng 117(3):417–423

Chieng CC, Launder BE (1980) On the calculation of turbulent heat transport downstream from an abrupt pipe expansion. Numer Heat Transf 3(2):189–207

Chiou CB, Wang CC, Lu DC (1995) Single-phase heat transfer and pressure drop characteristics of microfin tubes. ASHRAE Trans 101(2):1041–1048

Choi JY, Kedzierski MA, Domanski PA (2001) Generalized pressure drop correlation for evaporation and condensation in smooth and micro-fin tubes. In: Proceedings of IIF-IIR Commission B1, October 3–5, Paderborn, Germany, B4, pp 9–16

Chung BTF, Jia S (1995) A turbulent near-wall model on convective heat transfer from an abrupt expansion tube. Heat Mass Transf 31(1–2):33

Copetti JB, Macagnan MH, De Souza D, Oliveski De Cesaro R (2004) Experiments with microfin in single phase. Int J Refrig 27:876–883

DeLorenzo B, Anderson ED (1945) Heat transfer and pressure drop of liquids in double pipe fintube exchangers. ASME Trans 67:697–702

Eckels SJ, Pate MB (1991) An experimental comparison of evaporation and condensation heat transfer coefficients for HFC-134a and CFC-12. Int J Refrig 14:70–77

Edwards RJ, Jambunathan K, Button BL (1988) Experimental investigation of turbulent heat transfer in simultaneously developing flow in finned annuli. In: Shah RK, Ganic EN, Yang KT (eds) Proceedings of the first world conference on experimental heat transfer, fluid mechanics and thermodynamics. Elsevier Pub, pp 543–550

Esen EB, Obot NT, Rabas TJ (1994) Enhancement: part I. Heat transfer and pressure drop results for air flow through passages with spirally-shaped roughness. J Enhanc Heat Transf 1:145–156

Friedel L (1979) Improved friction pressure drop correlations for horizontal and vertical two-phase pipe flow, European Two-Phase Flow Group Meeting, paper E2

Fu WL, Wang CC, Chang WR, Chang CT (1995) Effect of anti-corrosion coating on the thermal characteristics of a louvered finned tube heat exchangers under dehumidifying conditions. In: Advances in enhanced heat/mass transfer and energy efficiency ASME, HTD, vol 320/PID, vol 1, pp 75–81

Gronnerud R (1979) Investigation of liquid hold-up, flow-resistance and heat transfer in circulation type evaporators, Part VI: two-phase flow resistance in boiling refrigerants, Bull. de l'Inst. du Froid

Gharebaghi M (2007) Numerical investigation of phase change process of PCM in a thermal energy storage system with fins. M.S. Thesis, Eastern Mediterranean University, Gazimagusa, North Cyprus

Gharebaghi M, Sezai I (2007) Enhancement of heat transfer in latent heat storage modules with internal fins. Numer Heat Transf Part A Appl 53(7):749–765

Gunter AY, Shaw WA (1942) Heat transfer, pressure drop, and fouling rates of liquids for continuous and noncontinuous longitudinal fins. ASME Trans 64:795–804

Haraguchi H, Koyama S, Esaki J, Fujii T (1993) Condensation heat transfer of refrigerants HCFC134a, HCFC123 and HCFC22 in a horizontal smooth tube and a horizontal microfin tube. In: Proceeding of 30th national symposium of Japan, Yokohama, Japan, pp 343–345, 43, Errata, p 791

Han DH, Lee KJ (2005) Single-phase heat transfer and flow characteristics of micro-fin tubes. Appl Therm Engg 25(11–12):1657–1669

Higahiiue S, Kuwahara K, Yanachi S, Koyama S (2007) Experimental investigation on heat transfer characteristics of supercritical carbon dioxide inside horizontal micro-fin copper tube during cooling process. In: ASME/JSME 2007 thermal engineering heat transfer summer conference collocated with the ASME 2007 InterPAC conference, American Society of Mechanical Engineers, pp 837–842

Hosseini MJ, Ranjbar AA, Sedighi K, Rahimi M (2012) A combined experimental and computational study on the melting behavior of a medium temperature phase change storage material inside shell and tube heat exchanger. Int Commun Heat Mass Transf 39(9):1416–1424

Hsieh WD, Chang KC (1996) Calculation of wall heat transfer in pipe expansion turbulent flows. Int J Heat Mass Transf 39(18):3813–3822

Huang XC, Ding GL, Hu HT, Zhu Y, Gao YF, Deng B (2010) Condensation heat transfer characteristics of R410A-oil mixture in 5 mm and 4 mm outside diameter horizontal microfin tubes. Exp Therm Fluid Sci 34:845–856

Hussein T, Shkarah AJ, Kazi SN, Badarudin A (2013) CFD simulation of heat transfer and turbulent fluid flow over a double forward-facing step. Math Probl Eng 2013:1–10

Hussein T, Ahmadi G, Abdulrazzaq T, Shkarah AJ, Kazi SN, Badarudin A (2015) Thermal performance of nanofluid in ducts with double forward-facing steps. J Taiwan Inst Chem Eng 47:28–42

Ito M, Kimura H (1979) Boiling heat transfer and pressure drop in internal spiral-grooved tubes. Bull JSME 22(171):1251–1257

Jensen MK, Vlakancic A (1999) Technical note–experimental investigation of turbulent heat transfer and fluid flow in internally finned tubes. Int J Heat Mass Transf 42:1343–1351

Ji WT, Zhang DC, He YL, Tao WQ (2012) Prediction of fully developed turbulent heat transfer of internal helically ribbed tubes—an extension of Gnielinski equation. Int J Heat Mass Transf 55 (4):1375–1384

Kedzierski MA, Goncalves JM (1999) Horizontal convective condensation of alternative refrigerants within a micro- fin tube. J Enhanc Heat Transf 6(1):161–178

Kelkar KM, Patankar SV (1990) Numerical prediction of fluid flow and heat transfer in a circular tube with longitudinal fins interrupted in the steamwise direction. J Heat Transf 112:342–348

Khanpara JC, Bergles AE, Pate MB (1986) Augmentation of R-113 in-tube condensation with micro fin tubes. In: Heat transfer in air conditioning and refrigeration equipment, HTD, vol 65. ASME, New York, pp 21–32

Khanpara JC, Bergles AE, Pate MB (1987) A comparison of in-tube evaporation of refrigerant 113 in electrically heated and fluid heated smooth and inner fin tubes. In: Jensen MK, Carey VP (eds) Advances in enhanced heat transfer, HTD, vol 68, pp 35–45

Khodadadi JM, Zhang Y (2001) Effects of buoyancy-driven convection on melting within spherical containers. Int J Heat Mass Transf 44:1605–1618

Koyama S, Kuwahara K, Fujii T, Inoue N, Hirakuni S (1993) Heat transfer and pressure drop of single phase flow inside internally grooved tubes. Naimen rasen mizotsuki kannai tansoryu no netsudentatsu oyobi atsuryoku sonshitsu

Koyama S, Yu J, Momoki S, Fujii T, Honda H (1996) Forced convective flow boiling heat transfer of pure refrigerants inside a horizontal microfin tube. In: Proceedings of the convective flow boiling, an international conference, pp 137–142

Kim DE, Kim MH (2010) Experimental study of the effects of flow acceleration and buoyancy on heat transfer in a supercritical fluid flow in a circular tube. Nucl Eng Des 240(10):3336–3349

Kim YJ, Jang J, Hrnjak PS, Kim MS (2009) Condensation heat transfer of carbon dioxide inside horizontal smooth and microfin tubes at low temperatures. J Heat Transf 131:1–10

Kwon JT, Park SK, Kim MH (2000) Enhanced effect of a horizontal micro-fin tube for condensation heat transfer with R22 and R410A. J Enhanc Heat Transf 7:97–107

Kuwahara K, Higashiiu S, Ito D, Koyama S (2012) Experimental study on cooling heat transfer of supercritical carbon dioxide inside horizontal micro-fin tubes. Trans Jpn Soc Refrig Air Condition Eng 24:173–181

Lacroix M, Benmadda M (1997) Numerical simulation of natural convection-dominated melting and solidification from a finned vertical wall. Numer Heat Transf A 31:71–86

Lam CKG, Bremhorst K (1981) A modified form of the k-ε model for predicting wall turbulence. J Fluids Eng 103(3):456–460

Launder BE, Sharma BI (1974) Application of the energy-dissipation model of turbulence to the calculation of flow near a spinning disc. Lett Heat Mass Transf 1(2):131–137

Lee HS, Kim HJ, Yoon J-I, Choi KH, Son CH (2013) The cooling heat transfer characteristics of the supercritical CO_2 in micro-fin tube. Heat Mass Transf 49(2):173–184

Li Y, Liu S (2013) Effects of different thermal conductivity enhancers on the thermal performance of two organic phase-change materials: paraffin wax RT42 and RT25. J Enhanc Heat Transf 20(6):463–473

Li W, Wu Z (2010a) A general criterion for evaporative heat transfer in micro/mini-channels. Int J Heat Mass Transf 53(1):1967–1976

Li W, Wu Z (2010b) A general correlation for adiabatic two-phase pressure drop in micro/mini-channels. Int J Heat Mass Transf 53:2732–2739

Li W, Wu Z (2010c) A general correlation for evaporative heat transfer in micro/mini-channels. Int J Heat Mass Transf 53(1):1778–1787

Li XW, Meng JA, Li ZX (2007) Experimental study of single-phase pressure drop and heat transfer in a micro-fin tube. Exp Thermal Fluid Sci 32(2):641–648

Li XW, Meng JA, Li ZX (2008) Enhancement mechanisms for single-phase turbulent heat transfer in micro-fin tubes. J Enhanc Heat Transf 15(3):227–242

Liu C, Groulx D (2014) Experimental study of the phase change heat transfer inside a horizontal cylindrical latent heat energy storage system. Int J Therm Sci 82:100–110

Liu XY, Jensen MK (2001) Geometry effects on turbulent flow and heat transfer in internally finned tubes. ASME J Heat Transf 123:1035–1044

Liu S, Huang Y, Liu G, Wang J, Leung LKH (2017) Improvement of buoyancy and acceleration parameters for forced and mixed convective heat transfer to supercritical fluids flowing in vertical tubes. Int J Heat Mass Transf 106:1144–1156

Manglik RM, Jog MA (2016) Resolving the energy-water nexus in large thermoelectric power plants: a case for application of enhanced heat transfer and high-performance thermal energy storage. J Enhanc Heat Transf 23(4):263–282

Mat S, Al-Abidi AA, Sopiana K, Sulaimana MY, Mohammada AT (2013) Enhance heat transfer for PCM melting in triplex tube with internal–external fins. Energy Convers Manage 74:223–236

Meyer JP, Olivier JA (2011a) Transitional flow inside enhanced tubes for fully developed and developing flow with different types of inlet disturbances: part I—adiabatic pressure drops. Int J Heat Mass Transf 54:1587–1597

Meyer JP, Olivier JA (2011b) Transitional flow inside enhanced tubes for fully developed and developing flow with different types of inlet disturbances: part II—heat transfer. Int J Heat Mass Transf 54:1598–1607

Moser KW, Webb RL, Na B (1998) A new equivalent Reynolds number model for condensation in smooth tubes. J Heat Transf 120(2):410–417

Mukkamala Y, Sundaresan R (2009) Single-phase flow pressure drop and heat transfer measurements in a horizontal microfin tube in the transition regime. J Enhanc Heat Transf 16 (2):141–159

Muller-Steinhagen H, Heck K (1986) A simple friction pressure drop correlation for two-phase flow pipes. Chem Eng Process 20:297–308

Murray RE, Groulx D (2014) Experimental study of the phase change and energy characteristics inside a cylindrical latent heat energy storage system: part 1 consecutive charging and discharging. Renew Energy 62:571–581

Myong HK (1990) A new approach to the improvement of k-ε turbulence model for wall-bounded shear flow. JSME Int J Eng 109:156–160

Nagano Y, Hishida M (1987) Improved form of the k-ε model for wall turbulent shear flows. J Fluids Eng 109(2):156–160

Nagano Y, Tagawa M (1990) An improved k-ε model for boundary layer flows. J Fluids Eng 112 (1):33–39

Narayanamurthy R (1999) Single phase turbulent flow in microfin and helically ribbed tubes. M.S Thesis, The Pennsylvania State University

Nuntaphan A, Kiatsiriroat T (2003) Heat transfer characteristic of cross flow heat exchanger using crimped spiral fin a case study of staggered arrangement. In: The 17th conference of mechanical engineering network of Thailand, Prachinburi Thailand

Nuntaphan A, Kiatsiriroat T, Wang CC (2005) Heat transfer and friction characteristics of crimped spiral finned heat exchangers with dehumidification. Appl Therm Eng 25(2–3):327–340

Ogoh W, Groulx D (2012) Effects of the number and distribution of fins on the storage characteristics of a cylindrical latent heat energy storage system: a numerical study. Heat Mass Transf 48 (10):1825–1835

Olivier JA, Liebenberg L, Thome JR, Meyer JP (2007) Heat transfer, pressure drop, and flow pattern recognition during condensation inside smooth, helical micro-fin, and herringbone tubes. Int J Refrig 30:609–623

Ooi TH, Webb DR, Heggs PJ (2004) A dataset of steam condensation over a double enhanced tube bundle under vacuum. Appl Therm Eng 24(8–9):1381–1393

Park KJ, Jung D (2008) Optimum fin density of low fin tubes for the condensers of building chillers with HCFC123. Energy Convers Manag 49(8):2090–2094

Rabas TJ, Eckels PW, Sabatino RA (1981) The effect of fin density on the heat transfer and pressure drop performance of low-finned tube banks. Chem Eng Commun 10(1–3):127–147

Raj R, Lakshman NS, Mukkamala Y (2015) Single phase flow heat transfer and pressure drop measurements in doubly enhanced tubes. Int J Therm Sci 88:215–227

Regin AF, Solanki SC, Saini JS (2008) Heat transfer characteristics of thermal energy storage system using PCM capsules: a review. Renew Sustain Energy Rev 12(9):2438–2458

Robinson KK, Briggs DE (1966) Pressure drop of air flowing across triangular pitch banks of finned tubes. Chem Eng Prog Symp Ser 62(64):177–184

Rozenfeld T, Kozak Y, Hayat R, Ziskind G (2015) Close-contact melting in a horizontal cylindrical enclosure with longitudinal plate fins: demonstration, modeling and application to thermal storage. Int J Heat Mass Transf 86:465–477

Saman W, Bruno F, Halawa E (2005) Thermal performance of PCM thermal storage unit for a roof integrated solar heating system. Sol Energy 78:341–349

Safaei MR, Togun H, Vafai K, Kazi SN, Badarudin A (2014) Investigation of heat transfer enhancement in a forward-facing contracting channel using FMWCNT nanofluids. Numer Heat Transf Part A Appl 66(12):1321–1340

Sapali SN, Patil PA (2010) Heat transfer during condensation of HFC-134a and R-404A inside of a horizontal smooth and micro-fin tube. Exp Therm Fluid Sci 34:1133–1141

Schlager LM, Pate MB, Bergles AE (1990) Evaporation and condensation heat transfer and pressure drop in horizontal 12.7-mm microfin tubes with refrigerant 22. J Heat Transf 112:1041–1047

Sciacovelli A, Gagliardi F, Verda V (2014) Maximization of performance of a PCM latent heat storage system with innovative fins. Appl Energy 137:707–715

Seeniraj RV, Velraj R, Narasimhan NL (2002) Thermal analysis of a finned-tube LHTS module for a solar dynamic power system. Heat Mass Transf 38(4–5):409–417

Shah MM (1979) A general correlation for heat transfer during film condensation inside pipes. Int J Heat Mass Transf 22:547–556

Sharifi N, Bergman TL, Faghri A (2011) Enhancement of PCM melting in enclosures with horizontally-finned internal surfaces. Int J Heat Mass Transf 54:4182–4192

Shedd TA, Newell TA (2003) Visualization of two-phase flow through microgrooved tubes for understanding enhanced heat transfer. Int J Heat Mass Transf 46(22):4169–4177

Shedd TA, Newell TA, Lee PK (2003) The effects of the number and angle of microgrooves on the liquid film in horizontal annular two-phase flow. Int J Heat Mass Transf 46(22):4179–4189

Taborek J (1997) Double-pipe and multitube heat exchangers with plain and longitudinal finned tubes. Heat Transf Eng 18(2):34–45

Tam LM, Ghajar AJ (1997) Effect of inlet geometry and heating on the fully developed friction factor in the transition region of a horizontal tube. Exp Therm Fluid Sci 15(1):52–64

Tam LM, Ghajar AJ (2006) Transitional heat transfer in plain horizontal tubes. Heat Transf Eng 27 (5):23–38

Tam HK, Tam LM, Ghajar AJ, Sun C, Leung HY (2011) Experimental investigation of the single-phase friction factor and heat transfer inside the horizontal internally micro-fin tubes in the transition region. In: Proceedings of ASME-JSME-KSME joint fluids engineering conference, Hamamatsu, Japan, 24–29 July 2011

Tam HK, Tam LM, Ghajar AJ, Tam SC, Zhang T (2012) Experimental investigation of heat transfer, friction factor, and optimal fin geometries for the internally microfin tubes in the transition and turbulent regions. J Enhanc Heat Transf 19(5):457–476

Thome JR (2004) Engineering Data Book III, Wolverine Tube, Inc. pp 68–79

Togun H, Abdulrazzaq T, Kazi SN, Kadhum AAH, Badarudin A, Ariffin MK, Sadeghinezhad E (2013) Numerical study of turbulent heat transfer in separated flow. Int Rev Mech Eng 7 (2):337–349

Togun H, Abu-Mulaweh HI, Kazi SN, Badarudin A (2016) Numerical simulation of heat transfer and separation Al_2O_3/nanofluid flow in concentric annular pipe. Int Commun Heat Mass Transf 71:108–117

Vakilaltojjar SM, Saman W (2001) Analysis and modelling of a phase change storage system for air conditioning applications. Appl Therm Eng 21:249–263

Voller VR (1990) Fast implicit finite-difference method for the analysis of phase change problems. Numer Heat Transf B 17:155–169

Wang HS, Rose JW (2004) Prediction of effective friction factors for single-phase flow in horizontal microfin tubes. Int J Refrig 27:904–913

Wang CC, Chiou CB, Lu DC (1996) Single-phase heat transfer and flow friction correlations for microfin tubes. Int J Heat Fluid Flow 17:500–508

Wang D, Tian R, Zhang Y, Li L, Shi L (2019) Experimental comparison of the heat transfer of supercritical R134a in a micro-fin tube and a smooth tube. Int J Heat Mass Transf 129:1194–1205

Webb RL, Scott MJ (1980) A parametric analysis of the performance of internally finned tubes for heat exchanger application. J Heat Transf 102:38

Webb RL, Narayanamurthy R, Thors P (2000) Heat transfer and friction characteristics of internal helical-rib roughness. J Heat Transf 122:134–142

Webb RL, Kim NH (2005) Principles of enhanced heat transfer, 2nd edn. Taylor & Francis, Boca Raton, FL

Wu Z, Li W (2011) A new predictive tool for saturated critical heat flux in micro/mini-channels: effect of the heated length-to-diameter ratio. Int J Heat Mass Transf 54:2880–2889

Yang CY (1999) A critical review of condensation heat transfer predicting models effect of surface-tension force. J Enhanc Heat Transf 6:217–236

Yu B, Feng Y, Tong L, Que X, Chen Z (1999) Theoretical analysis of fin efficiency with frost deposition on heat exchanger surface. In: Proceedings of the 5th ASME/JSME thermal engineering joint conference, Paper ATJE99-6402

Zalba B, Marin JM, Cabeza LF, Mehling H (2004) Free cooling of buildings with phase change materials. Int J Refrig 27:839–849

Zdaniuk G, Chamra LM, Mago PJ (2008) Experimental determination of heat transfer and friction in helically-finned tubes. Exp Thermal Fluid Sci 32(3):761–775

Zhang GM, Wu Z, Wang X, Li W (2012) Convective condensation of R410A in micro-fin tubes. J Enhanc Heat Transf 19(6):515–525

Chapter 8
Conclusions

We have discussed in detail externally finned tubes, internally finned tubes and annuli in this book. Following conclusions may be drawn:

- Wavy fins or some form of interrupted strip fins are used for gases.
- Analytical or numerical models are available to predict the heat transfer performance of high finned tube banks.
- Power law empirical correlations for plain fins have been developed.
- No general widely applicable correlation is available, although some correlations do predict data.
- Row effect of in-line and staggered tube arrangements makes the performance more complicated.
- In-line layouts have lower heat transfer performance than that by staggered tube layout.
- Both individually finned tube and plate fin-and-tube geometries are used for enhancement. Cost consideration plays a key role in decision-making.
- Higher performance can be obtained from oval or flat tubes.
- Hydrophilic coatings are used.
- Gas-side fouling limits the permissible enhanced fin geometry.
- Internally finned tubes and annuli are used to enhance tube-side heat transfer; many numerical analysis and empirically developed correlations are available.

© The Author(s), under exclusive license to Springer Nature Switzerland AG 2020   163
S. K. Saha et al., *Heat Transfer Enhancement in Externally Finned Tubes and Internally Finned Tubes and Annuli*, SpringerBriefs in Applied Sciences and Technology, https://doi.org/10.1007/978-3-030-20748-9_8

Additional References

Bae JH, Park M-H, Lee J-H (1999) Local flow and heat transfer of a 2-row offset strip fin-tube heat exchanger. J Enhanc Heat Transf 6:13–29

Balachandar S, Parker SJ (2002) Onset of vortex shedding in an in-line and staggered array of rectangular cylinders. Phys Fluids 14(10):3714–3732

Beecher DT, Fagan TJ (1987) Effects of fin pattern on the air-side heat transfer coefficient in plate finned-tube heat exchangers. ASHRAE Trans 93(2):1961–1984

Bergles AE (1985) Techniques to augment heat transfer. In: Handbook of heat transfer applications. McGraw-Hill, New York

Brauer H (1964) Compact heat exchangers. Chem Frog Eng (London) 45(8):451–460

Brigham BA, VanFossen GJ (1984) Length to-diameter ratio and row number effects in short pin-fin heat transfer. J Eng Gas Turbines Power 106:241–246

Chen L, Zhang HJ (1993) Convection heat transfer enhancement of oil in a circular tube with spiral spring inserts. In: Chow LC, Emery AF (eds) Heat transfer measurements and analysis, HTD-ASME Symp., vol 249, pp 45–50

Chokeman Y, Wongwises S (2005) Effect of fin pattern on the airside performance of herringbone wavy fin-and-tube heat exchangers. Heat Mass Transf 41:642–650

Chyu MK, Hsing YC, Natarajan V (1998) Convective heat transfer of cubic fin arrays in a narrow channel. J Turbomachinery Trans ASME 120:362–367 341

Chyu MK, Hsing YC, Shih TIP, Natarajan V (1999) Heat transfer contributions of pins and endwall in pin-fin arrays: effect of thermal boundary condition modeling. J Turbomachinery Trans ASME 121:257–263

Coetzee H, Liebenberg L, Meyer JP (2001) Heat transfer and pressure drop characteristics of angled spiralling tape inserts in a heat exchanger annulus. In: Paper RA, Aminemi NK, Toma O, Rudland R, Crain E (eds) Proceedings of the ASME process industries division. ASME, New York

Colburn AP, King WJ (1931) Relationship between heat transfer and pressure drop. Ind Eng Chem 23(8):918–923

Cox B, Jallouk PA (1973) Methods for evaluating the performance of compact heat exchanger surfaces. J Heat Transf 95:464–469

Date AW (1973) Flow in tubes containing twisted tapes. Heat Ventilating Eng 47:240–249

Date AW (1974) Prediction of fully-developed flow in a tube containing a twisted tape. Int J Heat Mass Transf 17:845–859

Date AW, Saha SK (1990) Numerical prediction of laminar flow in a tube fitted with regularly spaced twisted tape elements. Int J Heat Fluid Flow 11(4):346–354

© The Author(s), under exclusive license to Springer Nature Switzerland AG 2020
S. K. Saha et al., *Heat Transfer Enhancement in Externally Finned Tubes and Internally Finned Tubes and Annuli*, SpringerBriefs in Applied Sciences and Technology, https://doi.org/10.1007/978-3-030-20748-9

Date AW, Singham JR (1972) Numerical prediction of friction and heat transfer characteristics of fully developed laminar flow in tubes containing twisted tapes, ASME Paper 72-HT-1 7. ASME, New York

Davenport CJ (1984) Correlations for heat transfer and flow friction characteristics of louvered fin. In: Heat transfer-Seattle 1983, AIChE Symposium Series No. 225, vol 79, pp 19–27

Dhir VK, Chang F (1992) Heat transfer enhancement using tangential injection. ASHRAE Trans 98(2):383–390

Dhir VK, Tune VX, Chang F, Yu J (1989) Enhancement of forced convection heat transfer using single and multi-stage tangential injection. In: Goldstein RJ, Chow LC, Anderson EE (eds) Heat transfer in high energy heat flux applications, ASME Symp. HTD, vol 119

Ebisu T (1999) Development of new concept air-cooled heat exchanger for energy conservation of air-conditioning machine. In: Kakac S et al (eds) Heat transfer enhancement of heat exchangers. Kluwer Academic, Dordrecht, pp 601–620

Edwards FJ, Sheriff N (1961) The heat transfer and friction characteristics for forced convection air flow over a particular type of rough surface. In: International developments in heat transfer. ASME, New York, pp 415–425

Emerson WH (1961) Heat transfer in a duct in regions of separated flow. In: Proc. third international heat transfer conference, vol 1, pp 267–275

Engineering Sciences Data Unit. Low-fin staggered tube banks: heat transfer and pressure loss for turbulent single phase crossflow. Engineering Sciences Data Unit, ESDU Item 84016

Evans LB, Churchill SW (1963) The effect of axial promoters on heat transfer and pressure drop inside a tube, Chem. Eng. Prag. Symp. Ser: 59, vol 41, 36–46

Fiebig M, Mitra NK, Dong Y (1990) Simultaneous heat transfer enhancement and flow loss reduction of fin-tubes. Heat Transf 3:51–55

Fu WS, Tseng CC, Huang CS (1995) Experimental study of the heat transfer enhancement of an outer tube with an inner tube insertion. Int Heat Mass Transf 38(18):3443–3454

Gao X, Sunden B (2001) Heat transfer and pressure drop measurements in rib-roughened rectangular ducts. Exp Thermal Fluid Sci 24(1–2):25–34

Gianolio E, Cuti F (1981) Heat transfer coefficients and pressure drops for air coolers under induced and forced draft. Heat Transf Eng 3(1):38–48

Goldstein L, Sparrow EM (1976) Experiments on the transfer characteristics of a corrugated fin and tube heat exchanger configuration. J Heat Transf 98:26–34

Goldstein LJ, Sparrow EM (1977) Heat/mass transfer characteristics for flow in a corrugated wall channel. J Heat Transf 99:187–195

Gupte N, Date AW (1989) Friction and heat transfer characteristics of helical turbulent air flow in annuli. J Heat Transf 111:337–344

Hatada T, Senshu T (1984) Experimental study on heat transfer characteristics of convex louver fins for air conditioning heat exchangers, ASME paper, 84-H-74, New York

Hatada D, Ueda U, Oouchi T, Shimizu T (1989) Improved heat transfer performance of aircoolers by strip fins controlling air flow distribution. ASHRAE Trans 95(1):166–170

Hay N, West PD (1975) Heat transfer in free swirl in flow in a pipe. J Heat Transf 97:411–416

Hitachi Cable Ltd. (1984) Hitachi high-performance heat transfer tubes, Cat. No. EA-500. Hitachi Cable, Ltd, Tokyo, Japan

Holtzapple MT, Carranza RG (1990) Heat transfer and pressure drop of spined pipe in cross flow. Part 1: pressure drop studies. ASHRAE Trans 96(2):122–129

Holtzapple MT, Allen AL, Lin K (1990) Heat transfer and pressure drop of spined pipe in cross flow. Part III: heat transfer studies. ASHRAE Trans 96(2):130–135

Hong SW, Bergles AE (1976) Augmentation of laminar flow heat transfer in tubes by means of twisted tape inserts. J Heat Transf 98:251–256

Hong K, Webb RL (1999) Performance of dehumidifying heat exchangers with and without wetting coatings. J Heat Transf 121:1018–1026

Hong K, Webb RL (2000) Wetting coatings for dehumidifying heat exchangers. Int J Heat Ventilation Air Condition Refrig Res 6(3):229

Hou Kuan T, Lap Mou T, Ghajar AJ, Sik Chung T, Tong Z (2012) Experimental investigation of heat transfer, friction factor, and optimal fin geometries for the internally microfin tubes in the transition and turbulent regions. J Enhanc Heat Transf 19(5):457–476

Hsieh S-S, Kuo M-T (1994) An experimental investigation of the augmentation of tube-side heat transfer in a crossflow heat exchanger by means of strip-type inserts. J Heat Transf 116:381–390

Itoh M, Kogure H, Miyagi M, Mochizuki S, Yagi Y, Kunugi Y (1995) Development of an accordion-type offset fin heat exchanger. Trans Jpn Assoc Refrig 12(2):219–224. [in Japanese]

Jacobi AM, Shah RK (1998) Air-side flow and heat transfer in compact heat exchangers: a discussion of enhancement mechanisms. Heat Transf Eng 19(4):29–41

Jakob M (1938) Heat transfer and flow resistance in cross flow of gases over tube banks. Trans ASME 60:384

Jang J-Y, Chen L-K (1997) Numerical analysis of heat transfer and fluid flow in a three-dimensional wavy-fin and tube heat exchanger. Int J Heat Mass Transf 40:3981–3990

Jang JY, Yang JY (1998) Experimental and 3-D numerical analysis of the thermal hydraulic characteristics of elliptic finned-tube heat exchangers. Heat Transf Eng 19(4):55–67

Jang J-Y, Shieh K-P, Ay H (2001) Three-dimensional thermal-hydraulic analysis in convex louver finned-tube heat exchangers. ASHRAE Trans 107(2):503–509

Jayaraj D, Masilamani JG, Seetharamu KN (1989) Heat transfer augmentation by tube inserts in heat exchangers, SAE Technical paper 891983, Warrendale, PA

Jones TV, Russell CMB (1980) Heat transfer distribution on annular fins, ASME Paper 78-H-30, New York

Kang HC, Webb RL (1998) Evaluation of the wavy fin geometry used in air-cooled finned tube heat exchangers. In: Heat transfer 1998, Proceedings of the 11th international heat transfer conference Kyongju, Korea, vol 6, pp 95–100

Kim NH, Youn B (2013) Airside performance of fin-and-tube heat exchangers having sine wave or sine wave-slit fins. J Enhanc Heat Transf 20(1):43–58

Koch R (1958) Druckverlust und Waerrneuebergang bei verwirbeiter Stroemung, Vei: Dtsch. lngen Forschungsheft Ser. B 24(469): 1–44

Lopina RF, Bergles AE (1969) Heat transfer and pressure drop in tape-generated swirl flow of single-phase water. J Heat Transf 91:434–442

Maezawa S, Lock GSH (1978) Heat transfer inside a tube with a novel promoter. In: International Heat Transfer Conference Digital Library. Begel House Inc.

Manglik RM (1991) Heat transfer enhancement of in-tube flows in process heat exchangers by means of twisted-tape inserts. Ph.D. thesis, Department of Mechanical Engineering, Rensselaer Polytechnic Institute, Troy, NY

Manglik RM, Bergles AE (1992a) Heat transfer and pressure drop correlations for twisted-tape inserts in isothermal tubes: part I. Laminar flows. In: Pate MB, Jensen MK (eds) Enhanced heat transfer, ASME Symp. HTD, vol 202, pp 89–98

Manglik RM, Bergles AE (1992b) Heat transfer and pressure drop correlations for twisted-tape inserts in isothermal tubes: part II. Transition and turbulent flows. In: Pate MB, Jensen MK (eds) Enhanced heat transfer, ASME Symp. HTD, vol 202, 99–106

Matos RS, Laursen TA, Vargas JVC, Bejan A (2004) Three-dimensional optimization of staggered finned circular and elliptic tubes in forced convection. Int J Therm Sci 43:477–487

Mergerlin FE, Murphy RW, Bergles AE (1974) Augmentation of heat transfer in tubes by means of mesh and brush inserts. J Heat Trans 96:145–151

Metzger DE, Berry RA, Bronson JP (1982) Developing heat transfer in rectangular ducts with staggered arrays of short pin fins. J Heat Transf Trans ASME 104:700–706

Müller-Menzel T, Hecht T (1995) Plate-fin heat exchanger performance reduction in special two-phase flow conditions. Cryogenics 35(5):297–301

Oliver DR, Aldington RWJ (1986) Enhancement of laminar flow heat transfer using wire matrix turbulators. In: Heat transfer–1986 Proc. eighth international heat transfer conference, vol 6, pp 2897–2902

Park Y, Cha J, Kim M (2000) Heat transfer augmentation characteristics of various inserts in a heat exchanger tube. J Enhanc Heat Transf 7:23–34

Pulvirenti B, Matalone A, Barucca U (2010) Boiling heat transfer in narrow channels with offset strip fins: application to electronic chipsets cooling. Appl Therm Eng 30(14–15):2138–2145

Rabas TJ, Huber FV (1989) Row number effects on the heat transfer performance of inline finned tube banks. Heat Transf Eng 10(4):19–29

Razgatis R, Holman JP (1976) A survey of heat transfer in confined swirl flows. Heat Mass Transf Process 2:831–866

Rich DG (1973) The effect of fin spacing on the heat transfer and friction performance of multi-row, plate fin-and-tube heat exchangers. ASHRAE Trans 79(2):137–145

Rich DG (1975) The effect of the number of tube rows on heat transfer performance of smooth plate fin-and-tube heat exchangers. ASHRAE Trans 81(1):307–319

Robertson JM, Lovegrove PC (1983) Boiling heat transfer with Freon 11 brazed-aluminium plate fin heat exchanger. Trans ASME J Heat Transf 105:605

Saboya FEM, Sparrow EM (1976a) Transfer characteristics of two-row plate fin and tube heat exchanger configurations. Int J Heat Mass Transf 19:41–49

Saboya FEM, Sparrow EM (1976b) Experiments on a three-row fin and tube heat exchanger. J Heat Transf 98(3):520–522

Saboya FEM, Rosman EC, Carajilescov P (1984) Performance of one- and two-row tube and plate fin heat exchangers. J Heat Transf 106:627–632

Saboya SM, Saboya FEM (2001) Experiments on elliptic sections in one and two-row arrangements of plate fin and tube heat exchangers. Exp Therm Fluid Sci 24:67–75

Saha SK, Dutta A (2001) Thermohydraulic study of laminar swirl flow through a circular tube fitted with twisted tapes. J Heat Transf 123:417–427

Saha SK, Gaitonde UN, Date AW (1989) Heat transfer and pressure drop characteristics of laminar flow in a circular tube fitted with regularly spaced twisted-tape elements. Exp Therm Fluid Sci 2:310–322

Sahin B, Taguchi A (2007) Approach for determination of optimum design parameters for a heat exchanger having circular-cross sectional pin fins. Heat Mass Transf 43(5):493–502

Schmidt TE (1951) Heat transmission and pressure drop in banks of finned tubes and laminated coolers. Proc Gen Discuss Heat Transf II:186–188

Schmidt E (1963) Heat transfer at fumed tubes and computations of tube bank heat exchangers, Kaltetechnik, No. 4, 15, 98; No. 12, 15, 370

Seshimo Y, Fujii M (1991) An experimental study on the performance of plate and tube heat exchangers at low Reynolds numbers. In: Proc. 1981 ASME-JSME thermal engineering conference, vol 4, pp 449–454

Sethumadhavan R, Raja Rao M (1983) Turbulent flow heat transfer and fluid friction in helical wire coil inserted tubes. Int J Heat Mass Transf 26:1833–1845

Shah RK, Bhatti MS (1987) Laminar convective heat transfer in ducts. In: Kakac S, Shah RK, Aung W (eds) Handbook of single phase heat transfer. Wiley, New York, pp 3–20

Shivkumar C, Rao MR (1998) Studies on compound augmentation of laminar flow heat transfer to generalized power law fluids in spirally corrugated tubes by means of twisted tape inserts. In: Jacobs HR (ed) ASME Proc. 96, 1988 national heat transfer conference, HTD, 96, vol 1, 685–692

Smithberg E, Landis F (1964) Friction and forced convection heat transfer characteristics in tubes with twisted tape swirl generators. J Heat Transf 87:39–49

Thomas DG (1967) Enhancement of forced convection mass transfer coefficient using detached turbulence promoters. Ind Eng Chem Process Design Dev 6:385–390

Thorsen R, Landis F (1968) Friction and beat transfer characteristics in turbulent swirl flow subjected to large transverse temperature gradients. J Heat Transf 90:87–89

Torii K, Kwak KM, Nishina K (2002) Heat transfer enhancement accompanying pressure-loss reduction with winglet-type vortex generators for fin-tube heat exchangers. Int J Heat Mass Transf 45:3795–3801

Trupp AC, Lau ACY (1984) Fully developed laminar heat transfer in circular sector ducts with isothermal walls. J Heat Transf 106:467–469

Uttawar SB, Raja Rao M (1985) Augmentation of laminar flow beat transfer in tubes by means of wire coil inserts. J Heat Transf 105:930–935

Valencia A (1999) Heat transfer enhancement due to self-sustained oscillating transverse vortices in channels with periodically mounted rectangular bars. Int J Heat Mass Transf 42:2053–2062

Wang C-C, Chi K-Y, Chang C-J (2000a) Heat transfer and friction characteristics of plain fin-and-tube heat exchangers. Part II: correlation. Int J Heat Mass Transf 43:2693–2700

Wang C-C, Lin Y-T, Lee C-J (2000b) Heat and momentum transfer for compact louvered fin-and-tube heat exchangers in wet conditions. Int J Heat Mass Transf 43:3443–3452

Wang CC, Lee W-S, Sheu W-J (2001) A comparative study of compact enhanced fin-and-tube heat exchangers. Int J Heat Mass Transf 44:3565–3573

Wang CC, Liaw JS, Yang BC (2009) Airside performance of herringbone wavy fin-and-tube heat exchangers–data with larger diameter tube. Int J Heat Mass Transf 54:1024–1029

Webb RL (1983a) Enhancement for extended surface geometries used in air-cooled heat exchangers. In: Low Reynolds number flow heat exchangers. Hemisphere, Washington, DC, pp 721–734

Webb RL (1983b) Heat transfer and friction characteristics for finned tubes having plain fins. In: Low Reynolds number flow heat exchangers. Hemisphere, Washington, DC, pp 431–450

Webb RL (1990) Air-side heat transfer correlations for flat and wavy plate fin-and-tube geometries. ASHRAE Trans 96(2):445–449

Webb RL, Guptc N (1990) Design of light weight heat exchangers for air-to-two phase service. In: Shah RK, Kraus A, Metzger DE (eds) Compact heat exchangers: a Festschriftfor A.L. London. Hemisphere, Washington, DC, pp 311–334

Webb RL, Iyengar A (2000) Oval finned tube heat exchangers—limiting internal operating pressure. J Enhanc Heat Transf 8:147–158

Weierman C, Taborek J, Marner WJ (1978) Comparison of inline and staggered banks of tubes with segmented fins. AIChE Symp Ser 74(174):39–46

Wongwises S, Chokeman Y (2004) Effect of fin thickness on airside performance of herringbone wavy fin-and-tube heat exchangers. Heat Mass Transf 41:147–154

Xie L, Gu R, Zhang X (1992) A study of the optimum inserts for enhancing convective heat transfer of high viscosity fluid in a tube. In: Chen X-J, Veziroglu TN, Tien CL (eds) Multiphase flow and heat transfer; second international symposium, vol 1. Hemisphere, New York, pp 649–656

Yan W-M, Sheen P-J (2000) Heat transfer and friction characteristics of fin-and-tube heat exchangers. Int J Heat Mass Transf 43:1651–1659

Youn B (1997) Internal report, Samsung Electronic Co.

Youn B, Kil Y-H, Park H-Y, Yoo K-C, Kim Y-S (1998) Experimental study of pressure drop and heat transfer characteristics of 10.07 mm wave and wave-slit fin-tube heat exchangers with wave depth of 2 mm. In: Heat transfer 1998, Proceedings of 11th international heat transfer conference, vol 6, Kyongju, Korea

Youn B, Kim NH (2007) An experimental investigation on the airside performance of fin-and-tube heat exchangers having sinusoidal wave fins. Heat Mass Transf 43:1249–1262

Yun JY, Kim H-Y (1997) Structure of heat exchanger. U.S. patent 5697432

Yun JY, Lee K-S (2000) Influence of design parameters on the heat transfer and flow friction characteristics of the heat exchanger with slit fins. Int J Heat Mass Transf 43:2529–2539

Zhang YM, Han JC, Lee CP (1997) Heat transfer and friction characteristics of turbulent flow in circular tubes with twisted-tape inserts and axial interrupted ribs. J Enhanc Heat Transf 4 (4):297–308

Zhang LW, Balachandar S, Tafti DK, Najjar FM (1997a) Heat transfer enhancement mechanisms in in-line and staggered parallel-plate fin heat exchanger. Int J Heat Mass Transf 40(10):2307–2325

Zhang YM, Han JC, Lee CP (1997b) Heat transfer and friction characteristics of turbulent flow in circular tubes with twisted-tape inserts and axial interrupted ribs. J Enhanc Heat Transf 4 (4):297–308

Zhuo N, Ma QL, Zhang ZY, Sun JQ, He J (1992) Friction and heat transfer characteristics in a tube with a loose fitting twisted-tape insert. In: Chen X-J, Veziroglu TN, Tien CL (eds) Multiphase flow and heat transfer. Second international symposium, vol 1. Hemisphere, New York, pp 657–661

Zimparov V (2001) Enhancement of heat transfer by a combination of three-start spirally corrugated tubes with a twisted tape. Int J Heat Mass Transf 44:551–574

Zimparov V (2002) Enhancement of heat transfer by a combination of a single-start spirally corrugated tubes with a twisted tape. Exp Thermal Fluid Sci 25:535–546

Zimparov V (2004a) Prediction of friction factors and heat transfer coefficients for turbulent flow in corrugated tubes combined with twisted tape inserts. Part I: friction factors. Int J Heat Mass Transf 47:589–599

Zimparov V (2004b) Prediction of friction factors and heat transfer coefficients for turbulent flow in corrugated tubes combined with twisted tape inserts. Part 2: heat transfer coefficients. Int J Heat Mass Transf 47:385–393

Zukauskas A (1972) Heat transfer from tubes in cross flow. In: Hartnett JP, Irvine T Jr (eds) Advances in heat transfer, vol 8. Academic Press, San Diego, CA, pp 93–160

Index

A

Advanced internal fin geometries, 93
  condensation heat transfer coefficient,
    133, 136
  condensation pressure drop, 133, 135
  cross-flow heat exchanger, 141, 144
  cross-section view of tube, 153, 156
  efficiency index, 129, 130
  effectiveness, fins, 146, 149
  friction factor *vs.* Reynolds number, 153, 156
  heat transfer enhancement, 127
  heat exchangers, 150, 154
  kinetic energy, 150, 152
  mass flux
    vs. condensation frictional pressure
      drop, 133, 136
    vs. condensation heat transfer
      coefficient, 133, 136
  microfins (*see* Microfins)
  Nusselt number *vs.* Reynolds number, 153,
    156
  PCM (*see* Phase change material (PCM))
  pressure drop and heat transfer correlations,
    133, 135
  SIMPLE algorithm, 145, 147
  streamline of velocity, 150, 153
  surface heat transfer coefficient, 150, 151
  thermal energy, 143
  thermophysical properties, Al_2O_3
    nanoparticles and water, 150
  transient variations, melt fraction, 146, 149
  wall temperature and heat transfer
    coefficient variations, 138, 142
Aluminium fins, 143

Aluminium heat exchangers, 79
Annular winglet type VGs, 57
Annuli, 100, 108, 163
  axial fins, 137
  computational domains, 145, 146
  and correlations, 137
  double-pipe, 153
  finned cylindrical, 148
  PCM, 145
  plain cylindrical, 147
  thermal conductivity fluids, 153
  transient variations, melt fraction, 149
Arithmetic mean temperature difference
  (AMTD), 42

B

Blossom-shaped internal fins, 106
Buoyancy effect, 138

C

Circular fins
  effect of fin angles, 36
  fin spacing, 34
  finned tube banks, 34
  fin-tube heat exchanger, 31, 32
  longitudinal spacing, 31, 32
  radiation heat transfer, 36
  spiral and serrated fin geometry, 34–36
  staggered tubes, 31
  streamline plots, 32, 34
  temperature distribution, 32, 35
  transverse spacing, 31, 33

© The Author(s), under exclusive license to Springer Nature Switzerland AG 2020
S. K. Saha et al., *Heat Transfer Enhancement in Externally Finned Tubes and
Internally Finned Tubes and Annuli*, SpringerBriefs in Applied Sciences and
Technology, https://doi.org/10.1007/978-3-030-20748-9

Circular fins (*cont.*)
   tube bank geometries, 34
   VG, 31, 32, 34, 35
Circular grooved tube, 113
Coatings, 77–81
Colburn factor, 7
Convex louvre surface geometries, 48–50
Crimped spiral fins, 141, 144
Cross-flow heat exchangers, 141

D
Delta winglet type VGs, 58
Dimpled tubes
   geometrics, 97, 99
Double-pipe heat exchanger, 150

E
Entropy generation, 16, 18
Externally finned tubes, 163

F
Fanning friction factor, 7
Fin efficiency, 1, 2
Fin geometry, 16, 17
Fin numbering
   optimization, 16
   rate of entropy generation, 16, 18
   and spacing, 9
   total heat loss, 16, 19
Fin pitch
   axial, 46, 47
   circular, 46, 47
   heat transfer and pressure drop
      characteristics, 42
   independent, 52
   and tube rows, 48
Fin spacing, 144
   and development, 22
   four-row plain plate fin heat exchanger,
      18, 20
   vs. heat transfer coefficient, 23
   height ratio, 23, 24
   and numbers, 9
   plate cubic pin fin, 9
   thermal resistance, 13
   three-dimensional simulation, 22
Finned tube banks, 34
Finned tube heat exchangers
   air-side geometries, 1, 3
   circular tubes, 1, 2
   internally finned tubes, 2, 4

single-phase/two-phase heat transport, 1
   three-dimensional nature, 5
   three-row, 1, 4
   types, 2
   variables, 5
Finning disc and sample tube, 97, 98
Fin-tube heat exchanger, 31, 32, 48
Fitness-relative deviations, 88, 92
Flat microfin tubes
   cross section, 99, 100
   and oval (*see* Oval and flat tube geometries)
Friction factor, 152
   characteristics, 130
   definition, 5
   variation with Reynolds number, 11, 15

G
Gas-side fouling limits, 163
Gas-side heat transfer coefficient, 1
Global velocity distributions, 23
Goetler vortices, 40
Grooves, 112, 113, 115–118

H
HCFC123, 7
Heat exchanger geometries, 78
Heat-recovery equipment, 31
Heat transfer
   characteristics, 130
   efficiency, 115
   enhancement, 14
Helical fins, 85
Herringbone fin, 39, 40, 43
Hydrophilic coatings, 163

I
In-line arrangements, 22, 23, 25
In-line tube geometry, 18
Internal helical finned tube, 111, 112
Internally finned tubes, 2, 4, 163
   configurations, 94, 98, 106, 107
   dimensions, 85, 86
   experimental measurements, 95
   fitness-relative deviations, 88, 92
   friction data, 92, 93
   friction factor with Reynolds number, 85, 91
   fully developed laminar flow, 85, 89
   fully developed Nusselt number data, 92, 94
   geometrical parameters, 92
   hydrodynamic entrance length parameters,
      85, 90

in-line and staggered segmented internal
fins, 85, 89
natural convection, 85, 87
Nusselt number, 85–88
optimal fin number variation, 102, 106
optimization parameters, 102, 105
performance, 92, 96
Rayleigh number, 85, 86
segmented and continuous, 85, 90
solid model, 92, 96
S- , Z- and V-shaped fin profiles, 85, 91
thermal entrance lengths, 85, 90
thermal resistance, 102, 104
thermo-hydraulic performance, 102
three-dimensional, 93
tube geometries, 92, 97
values of different fluid–material, 85, 90
variation of Nusselt number, 85, 91
viscosity, 85, 88
wavy-fin array, 93, 97

J
Jacob correlation, 25

K
K-ε turbulence model, 22, 85
Kim correlation, 24
Kinetic energy, 150, 152

L
Laminar flow, 85–87, 89, 97, 111, 130,
150–152, 155
Large Eddy Simulation (LES) turbulence
model, 79
Local heat transfer coefficient
coatings, 77–81
numerical simulation, 77–81
patents, 77–81
performance comparison, 77–81
plain fins, 74, 77
Log mean temperature difference (LMTD), 42
Longitudinal spacing, 31–33
Louvred fin, 52, 54
automotive radiator geometry, 71
automotive radiator with in-line tubes, 73
geometric parameters, 8, 10
at low Reynolds number, 8
pressure drop *vs.* flow velocities, 8, 11
semi-louvred fin, 8, 11
and tube heat exchanger, 9
Low integral fin tube, 34

M
Mesh fin geometry, circular fin-tube heat
exchangers, 54–56
Microfins
buoyancy effect, 138
characteristics, 133
copper tube, 127
efficiency index, 138, 141
flow characteristics, 128
friction factors, 130, 133, 140, 143
and geometrical parameters, 139, 143
in heat exchanger, 139
heat transfer augmentation of single-phase
flow, 138
heat transfer characteristics, 130, 134, 139
heat transfer enhancement, 138, 139
helical tube, 127
horizontal, 150
inlet configurations, 130
isothermal friction enhancement index,
138, 140
laminar, transition and turbulent regions, 130
non-dimensional temperature t^+, 128
Nusselt number, 140
and plain tubes, 130, 131, 133
pressure drop, 139
R410A, 131
Reynolds number, 127, 141
sectional view, 130, 131
single-phase heat transfer enhancement,
127, 128
and smooth tube, 138
spiral angle, 130
tube geometry, 128, 129
two-phase heat transfer enhancement, 127
wall temperature difference, 139
Microfin tube, 85
aspect ratio, 101, 103
cross section, 99, 100
frictional pressure drops, 102, 104
geometrics, 99
heat transfer coefficients, 101
RMS errors, 101, 103
thermo-hydraulic performance of
R-410A, 97
transition of friction and heat transfer
data, 130, 132
Multi-louvred fins, 52
geometric parameters, 8

N
Non-dimensional parameters, 78
Numerical simulation, 77–81

O

Offset-strip fins, 52, 53
Optimal fin geometries
    characteristics, 130
OSF geometry, 51, 52
Oval and flat tube geometries
    air-side performance, 69, 71
    average Nu and f with fin pitch, 70, 73
    heat transfer and friction characteristics, 69, 70
    Nu_H vs. Re_H, 72, 76
    numerical calculation, 70, 71
    three-row staggered tube configuration, 72, 75
    tube dimensions, numerical calculations,
        69, 71
    VG, 72
    winglet pairs, 74

P

Paraffin wax, 143
Patents, 77–81
Perforated fin, 42, 52
Performance comparison, 77–81
Phase change material (PCM), 143–148
Pin-fin arrays, 58
Plain-plate fins
    average heat transfer coefficients, 19, 21
    average Nusselt number, 17, 20
    co-angular, 13
    Colburn factor, 7
    co-rotating pattern, 14
    Fanning friction factor, 7
    fin friction vs. Re_{st}, 19, 21
    flow pattern, 7
    friction factor, 22
    geometry, 9, 12
    global velocity distributions, 23
    heat transfer and friction characteristics,
        18, 20
    heat transfer coefficient, 9
        correlations, 11, 16
        and vapour quality, 7, 8
    heat transfer correlation, 19
    louvred fin (*see* Louvred fin)
    Nusselt number vs. Rayleigh numbers,
        7, 9, 14
    pressure drop, 22
    thermal resistance with fin spacing, 9, 13
    variation of η, 11, 16
    wavy fin (*see* Wavy fin)
Plain tube
    transition of friction and heat transfer
        data, 130, 132

Plate cubic pin fin
    geometry, 9, 12
Plate fin and circular fin geometries
    air-side test, 52, 54
    AMTD, 42
    convex louvre surface geometries, 48–50
    fin pattern, 50
    friction factor variation with Reynolds
        number, 39, 42, 50, 51
    heat transfer and pressure drop values, 43, 46
    herringbone fin (*see* Herringbone fin)
    LMTD, 42
    multilouvred fins, 52
    numerical results, 43, 45, 46
    Nusselt number and friction factor, 47
    Nusselt number variation, 39, 41, 42, 50, 51
    offset-strip fins, 52, 53
    one-row fin-tube heat exchanger, 52, 54
    OSF, 48, 51, 52
    perforated fin (*see* Perforated fin)
    performance evaluation index, Nu/f vs. Re,
        39, 42, 50, 52
    regression correlation, 40
    segmented/spine (*see* Segmented/spine fin)
    thermo-hydraulic characteristics, 42
    VGs (*see* Vortex generators (VGs))
Polyalphaolefin, 52
Pressure drop, 16, 152

R

Radiator, 18
Rectangular fins
    co-angular, zigzag, co-rotating and co-counter
        rotating configurations, 11, 15
Round tubes, plain-plate fins, *see* Plain-plate
        fins
Row correction factor, 34
Row effects
    in tube banks, 74, 76, 77

S

SAE5W30 engine oil, 52
Segmented/spine fin
    in air-conditioning applications, 60
    and circular fin geometries, 59
    copper tubes, 61
    seven-row in-line tube, 60
    staggered and in-line tube layouts, 60
Semi-implicit method for pressure-linked
        equations (SIMPLE) algorithm,
        145, 147

Semi-louvred fin, 8, 11
Smooth tube
    aspect ratio, 101, 103
    friction factor, 152
    frictional pressure drops, 102, 104
    geometrics, 99
    heat flux, 139
    heat transfer coefficients, 101
    mass fluxes, 138
    viscous sub-layer, 140
Smooth tube (SMT), 116
Smooth wave configuration, 39, 40, 47
Solid–liquid phase change process, 145
Spiral angle, 128
Spiral fin and serrated fin geometry, 34–36
Spirally fluted tubes
    aluminium, 107, 109
    axial distribution, heat transfer coefficient,
        112, 114
    curve fit and performance parameters,
        108, 110
    dimensionless geometric parameters,
        108, 109
    friction factor characteristics, 110
    indented tube, 107, 109
    in-tube flow and annular flow, 108
    mass flux effect, 101, 102
    Nusselt number vs. Darcy fiction factor,
        107, 108
    quality effect, 101, 102
    stainless steel tube, 107, 109
Staggered arrangement, 19, 22, 23
Staggered tubes, 31
Stainless steel, 116
Straight grooved tubes (SGT), 116
Surface temperatures, 16, 17

T
Taguchi method, 46
Test fins
    dimensions, 9, 12
Test section, 116, 117
Test tubes
    dimensions, 133, 134
    geometric parameters, 111, 128, 130,
        133, 134
    specifications, 131
Thermal energy, 143

Three-dimensional dimpled tubes
    heat transfer analysis, 95
    pressure drop analysis, 95
Transverse spacing, 31, 33
Tube banks, 5, 18, 22
    row effects, 74, 76, 77
Tube bundles
    dimensions, 22
    flow visualization, 55
Tube geometries, 92, 97
Turbulent flow, 85, 92, 93, 97, 106, 111, 112,
        127, 130, 133, 147, 150, 153

V
Vapour quality, 101
Vortex generator (VG), 14, 31, 32, 34, 35, 72
    annular winglet type, 57
    configurations, 43
    delta winglet type, 58
    fin-and-tube heat exchanger, 57
    fin, rib and wing configurations, 56
    fin surface, 55
    flow-down configuration, 56, 57
    flow-up configuration, 56
    heat exchanger, 56
    heat transfer rate and pressure drop, 43
    location, 43, 44
    louvre fin, 58
    and model names, 43, 44
    production and orientation, 58
    tube wake region, 59
    type, 43, 46
    winglet, 43, 58, 59
    wing-type, 56, 57

W
Wavy fin, 93, 97, 163
    behaviour, 13
    characteristic dimensions and top view, 7, 10
    geometric parameters, 8, 10, 39, 40
    heat transfer augmentation, 50
    heat transfer coefficient, 39
    smooth and herringbone, 39, 40
    thermo-hydraulic characteristics in
        air-cooled compact, 7
    thermo-hydraulic performance, 69
Wing-type VGs, 1, 43, 56, 57

Printed in the United States
By Bookmasters